Absolutely Mad Inventions

COMPILED
FROM THE RECORDS OF
THE UNITED STATES PATENT OFFICE
BY

A. E. BROWN
and H. A. JEFFCOTT, Jr.

DOVER PUBLICATIONS, INC.
NEW YORK

This Dover edition, first published in 1970, is an unabridged and unaltered republication of the work originally published by The Viking Press in 1932 under the title *Beware of Imitations!*

Standard Book Number: 486-22596-8
Library of Congress Catalog Card Number: 74-107666

Manufactured in the United States of America
Dover Publications, Inc.
180 Varick Street
New York, N.Y. 10014

Editor's Foreword

A study of the United States Patents reveals many curious and interesting devices from which those herein illustrated have been carefully selected. Space limitations have necessitated abbreviation of the text and, in some cases, of the drawings, but in each case care has been exercised in the selection in order that the patent may be truly and fairly represented. The drawings employed are exact copies of each of the selected illustrations, and those portions of the text which appear are direct quotations from the specifications.

In most instances it has been deemed advisable to append a title of our own in order that the purpose of the patent be quickly grasped, and in considering such patents we urge that the reader refer to the text for a complete understanding.

Should any reader wish more detailed information as to any of these patents, it is suggested that he write to the United States Patent Office at Washington, D. C. By giving the number of the patent and enclosing 50 cents, a complete copy of the original patent may be obtained.

The purpose of this book is not to criticize any patent or hold up to ridicule any inventor. We have no knowledge whatsoever as to whether any of the inventions have been commercial successes or failures. The reader, as we have done, can form an individual conclusion as to what commercial success each patent may have enjoyed. Far be it from us, who are neither inventors nor business men, to say that any of the patents referred to are not valuable inventions. In some instances, it may even be that, due to lack of finances or for some other unknown reason, the inventor has been unable to acquaint the general public with the nature and value of his invention. It is well known that the public has never had the opportunity of availing itself of many a worthy, practical, and helpful invention merely because of complete ignorance of its very existence.

It thus may be that through this little book we may bring to the inventor greater financial remuneration for his efforts and, at the same time, give the public information which will lead to so wide a use of the article as to have tremendous effect on the general mode of living.

Should this book be of exceptional benefit to any inventor or reader for the foregoing reasons, we should more than appreciate receiving letters advising us of that fact. Checks may even be enclosed if the individual feels financially indebted to us for our services.

ALFORD E. BROWN
HARRY A. JEFFCOTT, JR.

Contents

Absolutely Mad
Inventions

MEANS AND APPARATUS FOR PROPELLING AND GUIDING BALLOONS.

No. 363,037. Patented May 17, 1887.

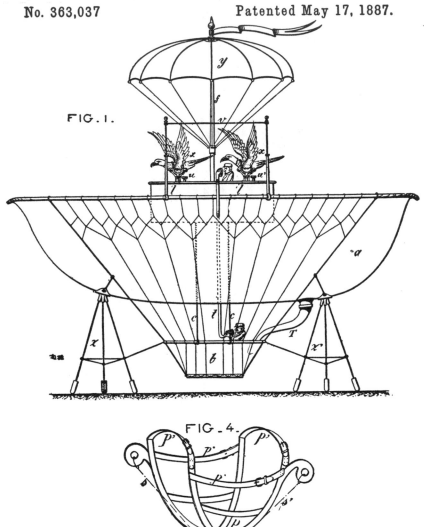

FIG.1.

FIG. 4.

Balloon Propelled by Eagles or Vultures

UNITED STATES PATENT OFFICE

MEANS AND APPARATUS FOR PROPELLING AND GUIDING BALLOONS

Specification forming part of Letters Patent No. 363,037, dated May 17, 1887
Application filed May 24, 1886. Serial No. 203,078. (No model.) Patented in France,
April 21, 1886, No. 175,662

. . . By this present invention the mechanical motor and propelling and guiding arrangements are replaced by a living motor or motors taken from the flying classes of birds—such as, for example, one or more eagles, vultures, condors, &c. By means of suitable arrangements (clearly shown in the annexed drawings) all the qualities and powers given by nature to these most perfect kinds of birds may be completely utilized. . . .

The corsets or harnesses p have forms and dimensions appropriate to the bodies of the birds chosen, such as eagles, vultures, condors, &c. . . .

It will now be understood that as the balloon floats in the air the man placed on the floor d can easily cause the cross k k' to turn by means of the hand-wheel m, and, with the cross, the birds x, so as to utilize their flight in the direction of the axis of the balloon, or in any other direction he desires. On the other hand, by working the rollers r r' he can direct the flight of the birds upward or downward. The result of these arrangements is that the flight of the harnessed birds must produce the motion and direction of the balloon desired by the conductor, whether for going forward or backward, in a right line, to the right or to the left, or for ascending or descending.

It may be observed that the birds have only to fly, the direction of their flight being changed by the conductor quite independently of their own will. . . .

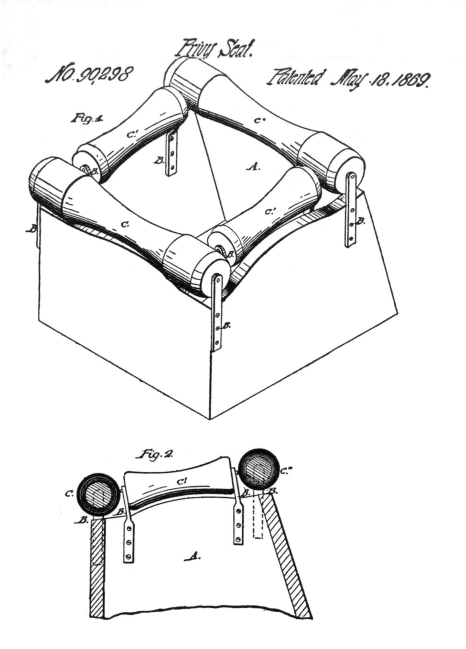

Privy Seat.

No. 90,298.

Patented May. 18. 1869.

Fig. 4.

Fig. 2.

UNITED STATES PATENT OFFICE

IMPROVEMENT IN PRIVY-SEATS

Letters Patent No. 90,298, dated May 18, 1869

. . . This invention relates to a device which renders it impossible for the user to stand upon the privy-seat; and consists in the provision of rollers on the top of the seat, which, although affording a secure and convenient seat, yet, in the event of an attempt to stand upon them, will revolve, and precipitate the user on to the floor. . . .

A represents the box, having one or more pairs of standards, B, which afford journal-bearing for a roller, C, over the front-edge of the box, and, where necessary, of side rollers C' C' and a back roller, C".

These rollers, while circular in transverse section, may have the represented or any other longitudinal contour, but are preferably somewhat hollowed toward their mid-length, as shown. . . .

EYE PROTECTOR FOR CHICKENS.

APPLICATION FILED DEC. 10, 1902.

NO MODEL.

UNITED STATES PATENT OFFICE

EYE-PROTECTOR FOR CHICKENS

No. 730,918   Patented June 16, 1903

Specification forming part of Letters Patent No. 730,918, dated June 16, 1903
Application filed December 10, 1902. Serial No. 134,679. (No model.)

. . . This invention relates to eye-protectors, and more particularly to eye-protectors designed for fowls, so that they may be protected from other fowls that might attempt to peck them, a further object of the invention being to provide a construction which may be easily and quickly applied and removed and which will not interfere with the sight of the fowl. . . .

1,175,513.

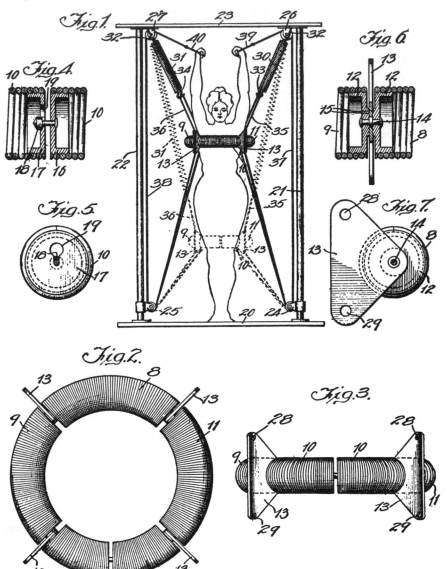

UNITED STATES PATENT OFFICE

MASSAGE APPARATUS

1,175,518   Specification of Letters Patent   Patented Mar. 14, 1916
Application filed June 8, 1914. Serial No. 842,556

... The primary object of my invention is to devise a tool which will mechanically simulate the rubbing or kneading action of a hand massage for the purpose of reducing flesh and accomplishing other remedial effects that a well executed manual massage is designed to accomplish.

In addition to the benefits obtained by the use of my improved massage tool, I have devised an apparatus for manually effecting vertical reciprocation of the tool whereby the user is not only benefited by the action of the tool itself, but by the exercise in connection with the operation of the tool as well.

Mechanical massage heretofore has been impracticable for the reason that where a tool has been used, it was almost impossible for the user to apply the tool to all parts of the body. This I accomplish mechanically by constructing a tool adapted to surround the body of the user and which is expansible, whereby the tool will conform to the contour of the body of the user in its reciprocating movement. . . .

The tool member shown in Fig. 2 may be made as large or as small as desired, according to the number of tool units that are coupled together. Where the body is to be massaged enough units are coupled together to surround the body of the user at its smallest circumference, for instance, at the waist. The tool may be opened by disconnecting the caps 16 and 17 so that it may be conveniently fixed in position around the body of the user whereupon the caps 16 and 17 are connected and the user thereupon grasps the handles 39 and 40. By moving the arms downwardly, the tool is pulled down along the body of the user, against the actions of the springs 30 and 31, until it reaches approximately the knees of the user, as shown in the diagrammatic position in Fig. 1.

In passing over the body of the operator, the tool expands or contracts, thereby automatically conforming to the exterior surface of the body, and as each unit comprising the tool is independently rotatable, and furthermore is capable of rotating on its own axis, the action of the tool simulates the rubbing or kneading action of a hand massage by the distorting and twisting action of the individual tool units in their rotating movement over uneven surfaces. . . .

RAILROAD TRAIN.

No. 536,360.　　　　　　　　Patented Mar. 26, 1895.

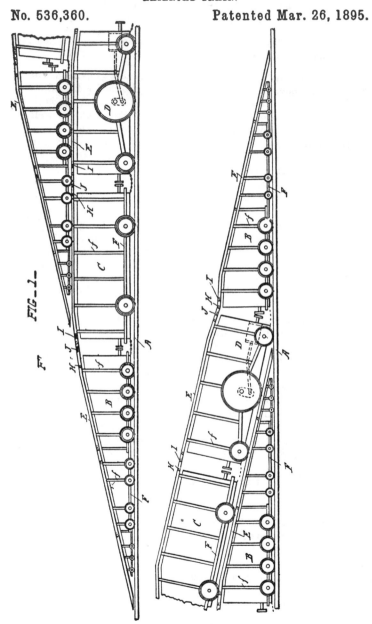

FIG_1_

Device to Prevent Train Collisions

UNITED STATES PATENT OFFICE

RAILROAD TRAIN

Specification forming part of Letters Patent No. 536,360, dated March 26, 1895
Application filed October 13, 1894. Serial No. 525,824. (No model)

. . . This invention relates to railroad trains; and it consists in the novel construction and combination of the parts hereinafter fully described and claimed, whereby one train may pass over another train which it meets or overtakes upon the same track. . . .

When one train meets or overtakes another train, one train will run up the rails E, carried by the other train, and will run along the rails E and descend onto the rails A at the other end of the lower train, as shown in Fig. 1.

The trains have the inclined lower ends of their rails E adjusted at different distances from the rails A in a prearranged manner so that the ends of the rails E of successive trains are not exactly upon the same level when they meet. The train having the ends of its rails E higher above the rails A than those of the train it meets will rise and run up on the rails E of the other train. . . .

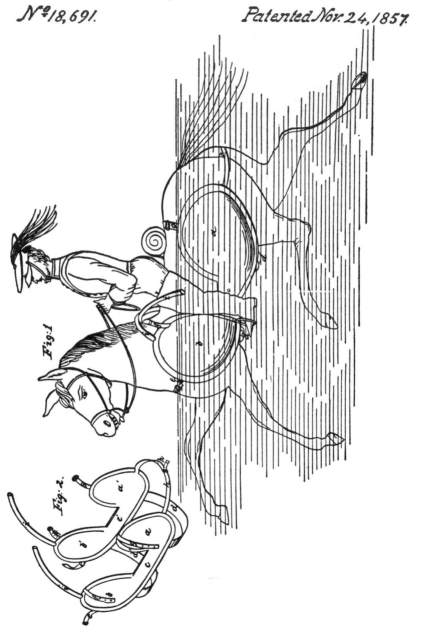

Equine Water-Wings

UNITED STATES PATENT OFFICE

IMPROVEMENT IN METHODS OF FLOATING HORSES, &c., ACROSS RIVERS

Specification forming part of Letters Patent No. 18,691, dated November 24, 1857

. . . During a long service on the frontier, where ferries are few, the necessity for some means more portable than boats for crossing the men and animals suggested itself. The want thereof often proves the source of much delay and loss, frequently the delay thus encountered in pursuing Indians being such as to make any further progress useless. Especially has this been experienced on the Pecos, in Texas, the Rio Grande, the Colorado and its tributaries, and on the Columbia and its tributaries. . . .

My floats consist, mainly, of a pair of bags of gutta-percha, india-rubber, or other suitable substance $a \ b \ c$ $a' \ b' \ c'$, each of the peculiar double-lobed form, substantially as represented, the two lobes $a \ b \ a' \ b'$ of each respective bag communicating interiorly by a small duct $c \ c'$. . . . i are tubes through which the bags may be inflated by the breath, the air thus introduced being secured by valves in any approved way. A squadron of cavalry thus equipped, the men having waterproof pantaloons with feet, can cross rivers, lakes, or estuaries dry-shod without the aid of boats. . . .

For emigrants in crossing the continent to California or Oregon, where numerous rivers are met with whose fords are doubtful or far asunder, a few such floats would prevent delay and the serious loss sometimes sustained by using the precarious alternative of rafts. . . .

23

DEVICE FOR SHAPING THE UPPER LIP

Filed March 25 , 1922

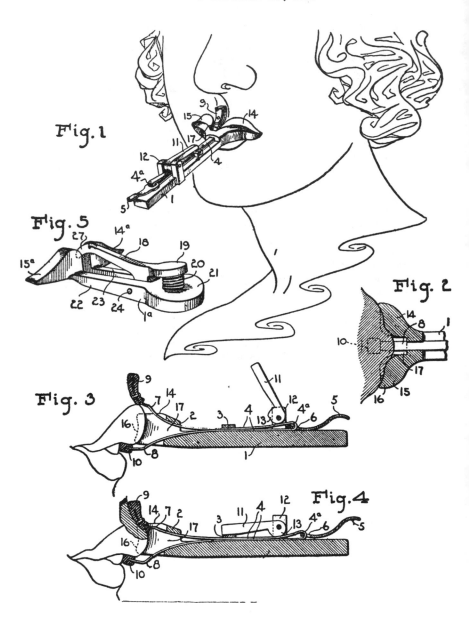

Fig.1

Fig. 5

Fig. 2

Fig. 3

Fig.4

Cupid's Bow Shaper

UNITED STATES PATENT OFFICE
DEVICE FOR SHAPING THE UPPER LIP

Patented June 10, 1924   1,497,842
Application filed March 25, 1922. Serial No. 546,846

. . . This invention relates to devices for re-shaping the upper lip of a person, and has for its object the provision of a simple and easily applied device to re-shape the upper lip of a person to conform to what is known as the "Cupid's bow," whereby it is unnecessary to resort to a surgical operation to produce this effect. . . .

By my new and improved device, I not only cause a depression to be formed in the upper surface and centrally of the upper lip, but the upper lip will be drawn into shaping relation with the matrix whereby the upper lip will eventually be changed to the form of the wellknown Cupid's bow. . . .

TIME ALARM.
APPLICATION FILED NOV. 16, 1907.

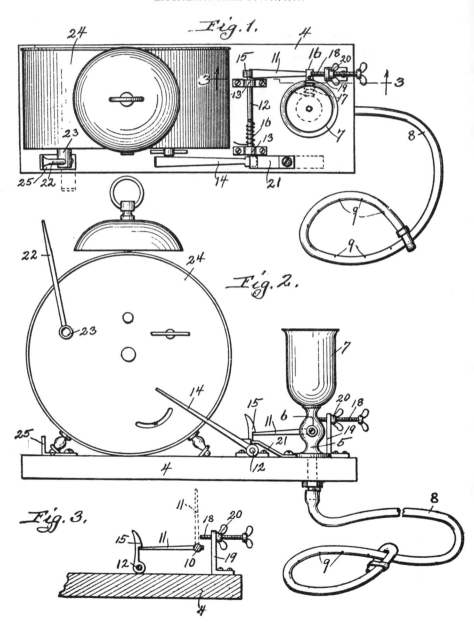

Fig.1.

Fig.2.

Fig.3.

Hydraulic Alarm Clock

UNITED STATES PATENT OFFICE

TIME-ALARM

No. 889,928   Specification of Letters Patent   Patented June 9, 1908

Application filed November 16, 1907. Serial No. 402,550

. . . My invention relates to alarms for waking a person out of sleep, and the objects of my invention are, first, to have an alarm sound at a predetermined time and the action of the alarm to open a valve which will permit water to flow on the person sleeping, second, to provide a perforated hose to direct water to the neck of a person; third to arrange mechanism whereby the alarm can be used alone; fourth, to make a device which will wake any person; fifth, to make a simple, cheap and durable device and other objects to become apparent from the description to follow: . . .

In operation the loop of the hose is placed about the neck of the person; the arm 11 is moved around until its free end is caught and retained by the catch 15; water is placed into the cup 7 and the clock alarm is set for the desired time in the usual manner. At the appointed time the clock alarm will sound, the winding stem turning in a reverse direction, *i. e.* clockwise as view in Fig. 2; the lever 22 will contact with and move the arm 11 a sufficient distance to release the catch 15 from the end of the arm 11; the spring 17 will then turn the valve 6 to an open position and the water from the cup 7 will flow through the hose 8 and out of the perforations 9. The lever 22 after moving the arm 11 will continue in its movement until it comes to stop against the bracket 25 pivotally secured to the board 4. . . .

DEVICE EMPLOYED FOR EXTERMINATING RATS, MICE, AND OTHER ANIMALS.

APPLICATION FILED DEC. 16, 1907.

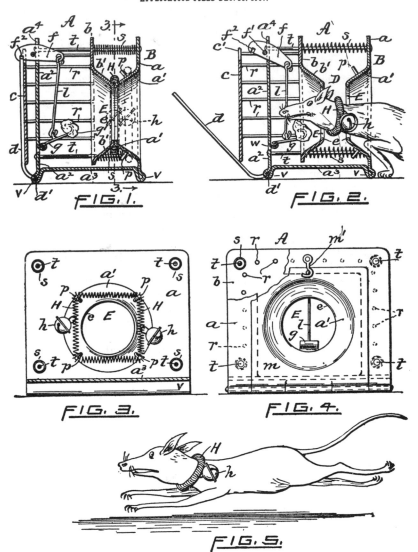

FIG. 1.

FIG. 2.

FIG. 3.

FIG. 4.

FIG. 5.

Humane Rodent Exterminator

UNITED STATES PATENT OFFICE

DEVICE EMPLOYED FOR EXTERMINATING RATS, MICE, AND OTHER
ANIMALS

No. 883,611   Specification of Letters Patent   Patented March 31, 1908
Application filed December 16, 1907. Serial No. 406,810

. . . Our invention relates to improved means for exterminating rats, mice, &c., and it consists essentially of a device
having a normally closed laterally separable annular frame,
an endless flexible resilient band or collar supported by and
encircling said annular part of the frame, and spring-
resisted tripping means operatively connected with a member of the frame, all constructed and arranged whereby
an animal, say a rat, upon introducing its head through the
frame opening and seizing the lure or bait attached to said
tripping means automatically releases the latter, which
action at the same instant also releases and separates the
said frame member and frees the expanded band, which
latter then immediately contracts around the animal's neck
before he can retreat from the device or apparatus. The
thus bedecked animal is not caught or confined in any
manner whatever but is free to return to its hole and colony.
The "bell-rat" as it may be termed, then in seeking its
burrow or colony announces his coming by the sounds
emitted by the bells, thereby frightening the other rats and
causing them to flee, thus practically exterminating them
in a sure and economical manner. It may be added that
the spring-band or collar is not liable to become accidentally
lost or slip from the rat's neck because the adjacent hairs
soon become interwoven with the convolutions of the spring
to more firmly hold it in place. . . .

1,045,060.

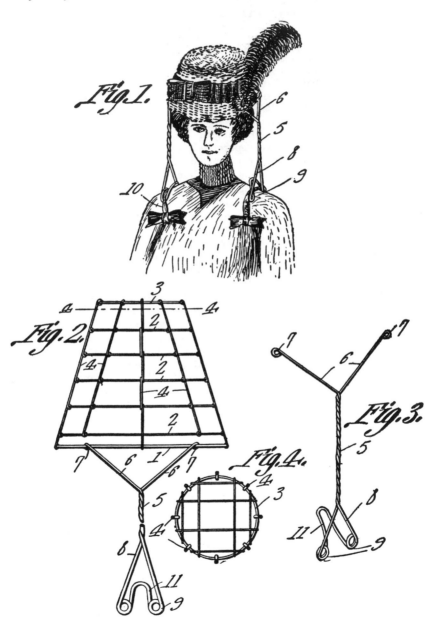

UNITED STATES PATENT OFFICE

HAT

1,045,060   Specification of Letters Patent   Patented Nov. 19, 1912
                Application filed May 10, 1911. Serial No. 626,232

. . . The objects of the invention are, to provide a hat which will permit of free circulation of air entirely around and over the head of the wearer, thus to prevent headaches caused by the weight and close fitting of the ordinary hat; to allow free movements of the head of the wearer independently of the hat; to afford unobstructed exhibition of the ornamentation and trimming of the wearer's hair and of the hat; to remove all weight from the head and transfer it to the shoulders of the user; to render it possible to employ a hat of such size as to avoid the use of a parasol or umbrella, and yet not in any way inconvenience the user by an added weight of material; to adapt a hat to be constructed of any material desired, such, for instance, as waterproof fabric, whereby to extend the range of its usefulness; to construct the article in such manner as to render it at once light, cheap and durable; and in general, to furnish a novel and thoroughly practical article of head-wear. . . .

For storm use, a rubber bag or covering may be employed, which may be placed over the exterior of the hat, or the frame itself may be covered with a waterproof material and thus provide an effective shield against moisture. . . .

1,046,177.

Patented Dec. 3, 1912.

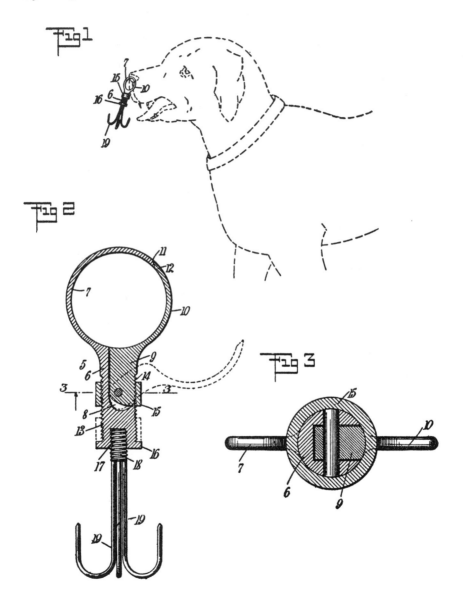

UNITED STATES PATENT OFFICE

DEVICE TO PREVENT DOGS FROM WORRYING SHEEP

1,046,177   Specification of Letters Patent   Patented Dec. 3, 1912

Application filed November 14, 1911. Serial No. 660,248

. . . My invention relates to devices to prevent dogs from worrying sheep, and it has for its object to provide one which may be fastened to the nose of a dog, and which is provided with hooks which will become entangled in the wool of a sheep so that when the sheep starts to run the dog's nose will be pulled, and the dog will receive a lesson which will break him of his habit of worrying sheep. . . .

When the hooks 19, with the stud 18, are secured to the ring member 5, which is attached to the nose of a dog, the hooks will become entangled in the wool of any sheep which the dog may attempt to worry. As soon as the dog is in close enough contact with the sheep to permit the hooks 19 to become entangled in the wool of the sheep, the sheep will start to run, which will yank the ring member 5, and give the dog's nose a severe pull, and, after a few attempts have been made in this way to worry the sheep, it will be found that the dog is very careful not to go too close to the sheep and that the sheep are no longer disturbed at the presence of the dog. . . .

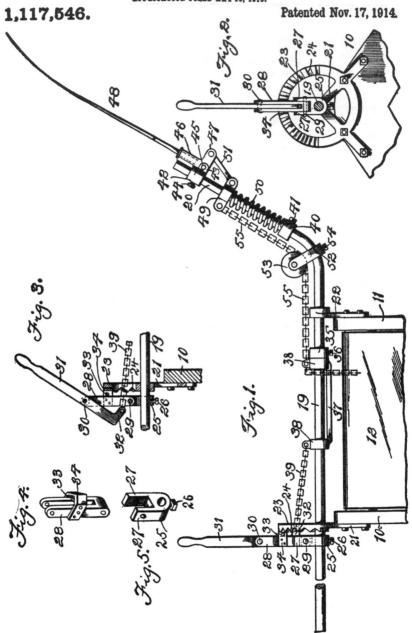

Labor-Saving Whip

UNITED STATES PATENT OFFICE

MECHANICAL WHIP

1,117,546   Specification of Letters Patent   Patented Nov. 17, 1914
          Application filed May 16, 1913. Serial No. 768,179

. . . This invention relates to mechanical whips for attachment to a vehicle to enable the driver to apply a whip to any horse in a team, and has for one of its objects to improve the construction and increase the efficiency and utility of devices of this character. . . .

The operator rotates the member 19 until the oblique extension 20 is at the proper angle. He then moves the upper end of the lever 31 forwardly upon its pivot 30 which movement exerts a pulling force upon the flexible member 39 and moves the frame 37—38 longitudinally of the member 19 which movement causes the flexible member 55 to move the sleeve 49 downwardly and thus through its connection by the link 51 with the whip holder 46 moves the whip downwardly and applies it to the horse. When the lever section 31 is released the reaction of the spring 50 restores the whip to its upper position. By this simple arrangement it will be obvious that the driver from his seat can adjust the whip to any desired position and apply the same to any horse of the team, the desired movement being accomplished by the one lever.

When two teams are employed, one in advance of the other, the horizontal section 19 will be adjusted to its farthest point to bring the whip supporting portion opposite the forward team, and then by adjusting the members 19—20 rotatively, as above described, the whip may be applied to either one of the horses of the forward team, the chain 55 being of sufficient length to enable it to be adjusted to correspond to the necessary adjustments of the parts, as will be obvious. By this simple arrangement it will be obvious that the position of the whip is under the complete control of the driver on the seat, and by its use he is enabled to apply the whip to any horse in the team, no matter how many horses may be employed, or no matter how arranged, whether all of the horses are abreast, or arranged with one team ahead of the other. The horizontal section 19 of the device will be increased or decreased in length to correspond to the construction of the vehicle to which it is attached and to correspond to the number of horses employed in connection therewith. . . .

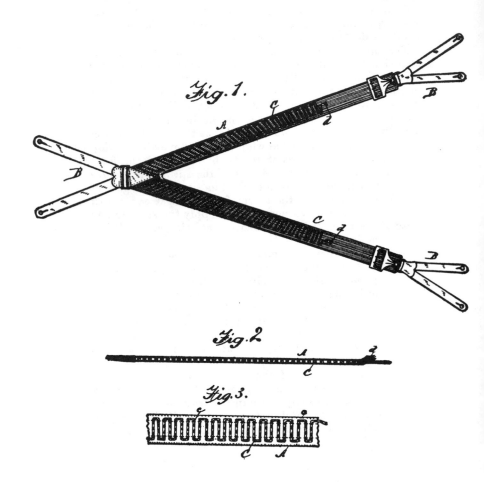

Fig.1.

Fig.2.

Fig.3.

UNITED STATES PATENT OFFICE
SUSPENDERS
Specification forming part of Letters Patent No. 323,416, dated August 4, 1885
Application filed February 3, 1885. (No model)

. . . My invention relates to improvements in suspenders, having for its object to provide a suspender with a cord so secured thereto or formed therewith as to constitute a part of the same, and to be readily and easily detached therefrom, whereby, in the event of a person being confined to a burning building and having all of the usual means of escape cut off, the cords can be disengaged from the suspenders and lowered to the ground to receive a rope, and thus enable the person to effect his escape. . . .

ELECTRIC EXTRACTION OF POISONS.

(Application filed Oct. 5, 1896.)

(No Model.)

Electrotherapeutic Device

UNITED STATES PATENT OFFICE

ELECTRIC EXTRACTION OF POISONS

Specification forming part of Letters Patent No. 606,887, dated July 5, 1898
Application filed October 5, 1896. Serial No. 607,955. (No model)

... Be it known that I ... have invented certain new and useful Improvements in the Electrical Extraction of Poisons from the Human Body; and I hereby declare that the accompanying is a full, clear, and exact description of the same, reference being had to the accompanying drawing, in which the figure is a view in perspective of a male subject or patient seated in a chair, the electric battery, and the conducting-wires leading from the electric battery to the positive and negative plates, which in the illustration are shown applied to the back of the neck of the patient and at the same time to the bare feet of the patient or person receiving treatment. . . .

For vegetable poisons I employ a vegetable receiver instead of a mineral or copper one, and for animal poisons I use an animal receiver, such as raw meat, the device being capable of use with the mineral, vegetable, or animal receivers without further change than to equip it with the kind of receiver applicable to the kind of poison desired to be extracted or removed from the human system. . . .

The application of the different receivers is made to the negative electrode, and the positive electrode is applied to any suitable part of the body. When the current is turned on, it will run down from the neck or other suitable place through the patient's body and will pull or draw out the poison at the negative pole and deposit it on the copper plate. From six to eight treatments of a half an hour each in duration will generally extract all of the poison of whatever kind it may be, and the copper plate will show as bright and clear as it was at first. The copper plate or other receiver may be applied to any part of the human body where poison may be found. . . .

BADGE.

No. 174,162. Patented Feb. 29, 1876.

Fig. 1

Fig. 2

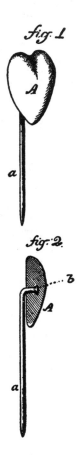

UNITED STATES PATENT OFFICE

IMPROVEMENT IN BADGES

Specification forming part of Letters Patent No. 174,162, dated February 29, 1876
Application filed January 17, 1876

. . . This invention relates to an improvement in articles of confectionery, the object being to form a badge of confectionery combined with a pin, by which it may be attached to the garment; and it consists in a pin with the badge or article formed from confectionery, cast upon and around the said head, as hereinafter described. . . .

This in no way destroys the confectionery, and may be disposed of in the usual manner for confectionery. . . .

922,956.

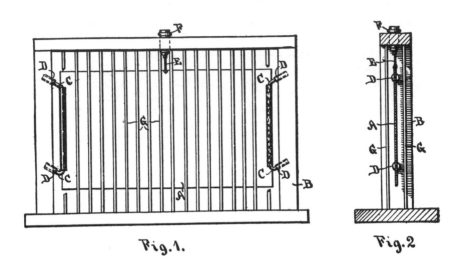

Fig.1.

Fig.2

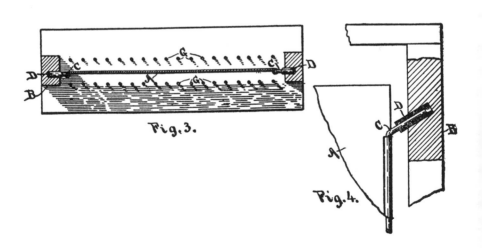

Fig.3.

Fig.4.

Electrical Nuisance Preventer

UNITED STATES PATENT OFFICE

DEVICE FOR PREVENTING DOG NUISANCE

No. 922,956   Specification of Letters Patent   Patented May 25, 1909

Application filed December 26, 1907. Serial No. 408,190

. . . This invention relates to improvements in devices for preventing dogs from committing the nuisance of urinating against buildings, walls, and other structures; and my object is to provide a device which may be conveniently placed at points where such nuisance has been, or is likely to be, committed, whereby an electric shock will be administered to a dog, when attempting to commit such nuisance, which will effectually prevent a recurrence of the act in that locality by the dogs so punished. . . .

The plate, when coupled to an electric light circuit, or to a battery or other source of electric current of suitable power, becomes a terminal from which no current will pass until a ground connection is made. When, therefore, the device has been placed in position in front of a building, or the like, where dogs have been in the habit of committing nuisance, or are likely to commit nuisance, the next dog that attempts the act, will receive a severe shock the instant the stream of urine strikes the plate, by reason of the grounding of the current through the dog's body. After receiving one such shock it is believed that that particular locality will be shunned in the future by every dog so punished. . . .

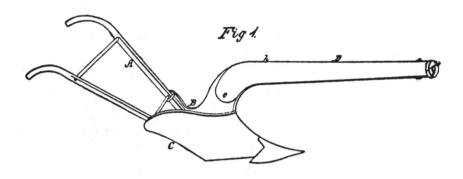

Fig. 1.

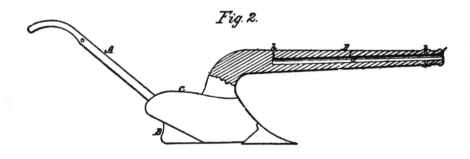

Fig. 2.

UNITED STATES PATENT OFFICE

IMPROVEMENT IN COMBINED PLOW AND GUN

Specification forming part of Letters Patent No. 35,600, dated June 17, 1862

. . . The object of our invention is to produce a plow equal, if not superior, in point of strength and lightness to that implement as ordinarily made, and at the same time to combine in its construction the elements of light ordnance, so that when the occasion offers it may do valuable service in the capacity of both implements. . . .

It is symmetrical and pleasing to the eye. As a piece of light ordnance its capacity may vary from a projectile of one to three pounds weight without rendering it cumbersome as a plow. Its utility as an implement of the twofold capacity described is unquestionable, especially when used in border localities, subject to savage feuds and guerrilla warfare. As a means of defense in repelling surprises and skirmishing attacks on those engaged in a peaceful avocation it is unrivaled, as it can be immediately brought into action by disengaging the team, and in times of danger may be used in the field, ready charged with its deadly missiles of ball or grape. The share serves to anchor it firmly in the ground and enables it to resist the recoil, while the hand-levers A furnish convenient means of giving it the proper direction.

This combination enables those in agricultural pursuits to have at hand an efficient weapon of defense at a very slight expense in addition to that of a common and indispensable implement, and one that is hardly inferior as regards the means of moving, planting, and directing to that of expensive light ordnance on wheels. . . .

COMBINED GROCER'S PACKAGE, GRATER, SLICER, AND MOUSE AND FLY TRAP.

No. 586,025.

Patented July 6, 1897.

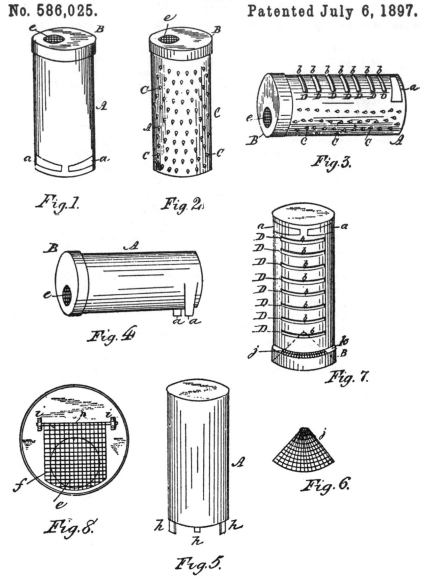

Fig. 1.

Fig. 2.

Fig. 3.

Fig. 4.

Fig. 7.

Fig. 8.

Fig. 5.

Fig. 6.

UNITED STATES PATENT OFFICE

COMBINED GROCER'S PACKAGE, GRATER, SLICER, AND MOUSE AND FLY TRAP

Specification forming part of Letters Patent No. 586,025, dated July 6, 1897

Application filed January 13, 1897. Serial No. 619,032. (No model)

. . . My invention relates to the construction of a grocer's sheet-metal box in such a manner as to be useful for other purposes, (after the first contents are removed,) such as a grater, shredder, slicer, and mouse and fly trap. I attain these objects by the construction illustrated in the accompanying drawings, in which—

Figure 1 represents a sheet-metal cylindrically-shaped grocer's package or box with feet attached to one end. Fig. 2 represents a similar figure, showing the opposite side constructed as a grater. Fig. 3 represents the same constructed as a slicer. Fig. 4 represents the same as a mouse-trap. Fig. 5 represents the body with vertical feet at one end, preparatory to using it as a fly-trap. Fig. 6 represents the conical wire attachment for the box when used as a fly-trap. Fig. 7 represents the box, the wire-gauze, and box-cover used, complete for a fly-trap. . . .

On the lid B of the box a circular hole e is cut about an inch and a quarter in diameter, and on the inside of the said cover is hinged a wire door f at the top by the upper wire n passing through hinge-plates i i, soldered to the lid, as shown at Fig. 8. The mice will enter the opening e. The wire door f, swinging inward, will enable them to do so, and when they pass it it drops down against the said opening and closes it so effectually as to prevent all egress of the mice.

When the box A is to be used as a fly-trap, a cone-shaped wire diaphragm j, with a hole in the center, as in Fig. 6, is placed at the lid end of the box A when it is laid on a table or other convenient place, three strips of sheet metal h previously being soldered to the mouth of the box A, and when the box is used as a grocer's package they are bent back out of the way, and when used as a fly-trap the said strips are bent outward, projecting from the mouth of the box about half an inch, and when the box is inverted on the cover the said strips form legs, upon which the box stands, leaving an annular space k of about one-quarter of an inch wide between the edge of the box A and its inverted cover B beneath it for the ingress of flies, who enter and crawl up the wire cone through a hole in the apex and are encaged in the upper part of the said box A. . . .

Rocking Chair.

No. 92,379.   Patented July 6.1869,

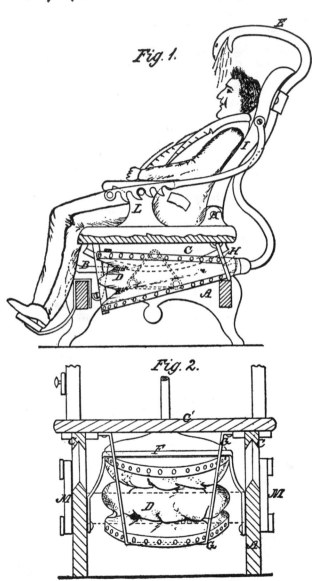

Fig. 1.

Fig. 2.

UNITED STATES PATENT OFFICE

IMPROVED ROCKING-CHAIR

Letters Patent No. 92,379, dated July 6, 1869

The Schedule referred to in these Letters Patent and making part of the same.

. . . This invention relates to improvements in the construction of rocking-chairs, with air-blowing attachments, having for its object to provide a stand or base for the support of a bellows, with tracks or rails, on which the rockers, which are fixed close to the seat, may work, instead of on the floor; also, to provide an arrangement whereby the parts may be readily detached for storage or packing in compact form; and also an improved arrangement of parts, whereby the bellows is operated, all as hereinafter specified. . . .

This stand, with elevated rails, protects the rockers against rocking on small children crawling on the floor, or strings scattered thereon. It also provides for rocking the chair with the same ease, when sitting on the ground; and it also serves as a support for a bellows, D, whereby the occupant may, by the act of rocking, impel a current of air upon himself, through a flexible tube, E, which may be directed to any part, as required.

The top of this bellows is connected by a bent bar, F, to the stand A, so as to be held in a fixed position, while the lower part is connected by a similar bent bar, G, to the bottom of the chair, so as to be moved up and down with it, to impel the air. . . .

CHEWING GUM LOCKET.

No. 395,515.   Patented Jan. 1, 1889

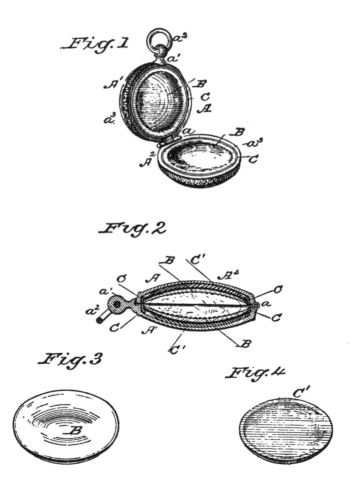

Fig. 1

Fig. 2

Fig. 3

Fig. 4

UNITED STATES PATENT OFFICE

CHEWING-GUM LOCKET

Specification forming part of Letters Patent No. 395,515, dated January 1, 1889
Application filed September 10, 1887. Serial No. 249,394. (No model)

. . . The object of my invention is to provide a locket of novel form and construction for holding with safety, cleanliness, and convenience for use chewing-gum, confections, or medicines, and which may be carried in the pocket or otherwise attached to the person, as lockets are ordinarily worn; and the improvement consists, essentially, in a locket having an anti-corrosive lining, and it also consists in certain details of construction and combinations of parts, hereinafter particularly described, and designated in the claims. . . .

As the lining B is made of a non-corrosive material, any of which may be employed without departing from my invention, the saliva of the mouth or other substance held within the locket will not act upon it chemically, and a case of any preferred material may thus be used. Chewing-gum may thus be carried conveniently upon the person, and is not left around carelessly to become dirty or to fall in the hands of persons to whom it does not belong, and be used by ulcerous or diseased mouths, by which infection would be communicated by subsequent use to the owner. . . .

(No Model.)

ILLUMINATING DEVICE FOR FRIGHTENING RATS AND MICE.

No. 305,102.   Patented Sept. 16, 1884.

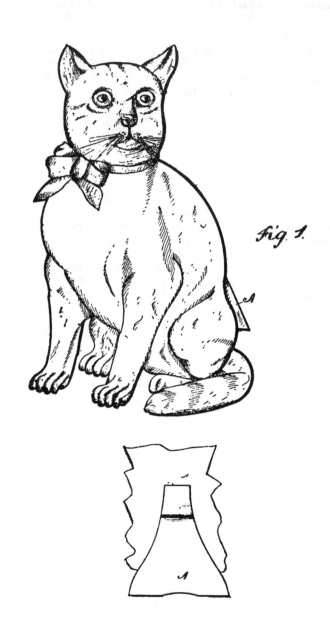

Fig. 1.

Ornamental Rat Exterminator

UNITED STATES PATENT OFFICE

ILLUMINATING DEVICE FOR FRIGHTENING RATS AND MICE

Specification forming part of Letters Patent No. 305,102, dated September 16, 1884
Application filed July 30, 1884. (No model)

. . . This invention relates to a new, useful, and ornamental illuminated device for frightening and exterminating rats and mice; and it has for its object to provide an article of this character which will be arranged and adapted to effect the purposes stated without the use of deadly poisons.

To this end the said invention consists in printing the figure of a cat on card-board having several coats of illuminating-paint arranged so that the figure will shine in the dark; and, furthermore, in perfuming said figure with peppermint, which is obnoxious to rats, and mice, and thus the device will have the effect to drive away these rodents. . . .

Referring to the drawings, it will be seen that I have shown the figure of a cat cut out of card-board and painted to present an attractive appearance, the cat being shown in a sitting posture, with its head turned toward the right and its eyes directed toward and watching an object near by. Over this painted figure I apply several coats of illuminating-paint, so that it will shine in the dark, and then I perfume the figure with oil of peppermint, which is obnoxious to rats and mice, and will serve as an exterminator. The eyes of the cat are coated with a thick coat of phosphorus, so as to shine out with more brilliance than the body of the figure. To the back of the figure is attached a swinging flap, A, arranged to be folded flat against the back or swung outward to rest on the stand or floor, so as to support the figure in an upright position. . . .

As a parlor-ornament the device serves two functions, since it will frighten away rats and mice, and forms a useful and attractive article to place on the mantel-piece or stand. It is also useful to place on the window-sill facing the window, so as to shine through the same and be seen in the dark. It can also be placed in the pantry, on the shelves or floor adjacent to the rat-hole, or near the parts traversed by the rats or mice, and by the peculiar but not offensive odor with which the figure is permeated it will act as an effectual exterminator. . . .

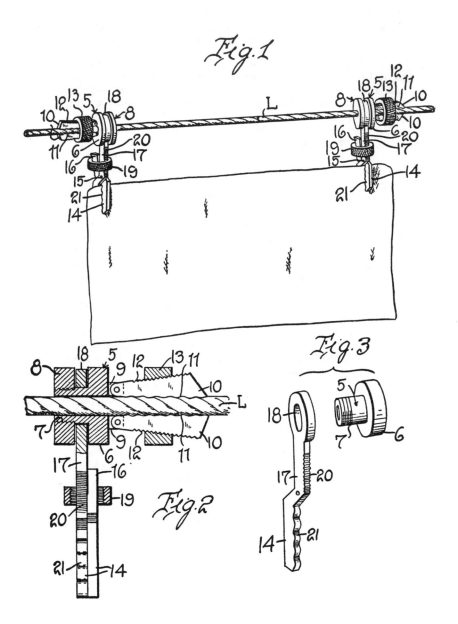

Fig.1

Fig.2

Fig.3

Mechanical Clothes-Pin

UNITED STATES PATENT OFFICE

CLOTHES-PIN

1,159,804   Specification of Letters Patent   Patented Nov. 9, 1915
Application filed May 19, 1915. Serial No. 29,141

. . . This invention relates to an improved clothes pin or fastener and has for its primary object to provide a very simple device of this character which will serve to securely hold the articles upon the line and yet permit of their swinging or swaying movement with respect thereto when blown by high winds. . . .

In the use of my invention, the line indicated at L is passed through the central bore or opening of the body member 5, the eye 18 of the clothes fastener being engaged upon said body against one side of the flange 5. It will be understood that the nut 13 is threaded inwardly upon the dogs 10 so that said dogs may be spaced sufficiently to permit of the free sliding movement of the clamp upon the line. The ring 8 is threaded upon the end 7 of the clamp body and the several parts are then moved to the desired position upon the clothes line. The nut 13 is now threaded outwardly upon the wider ends of the dogs 10 so that said dogs will be forced inwardly and their toothed faces 11 caused to securely grip on opposite sides of the line L. The article is now disposed between the jaws 14 and the nut 19 threaded upwardly upon the shanks 16 and 17 of said jaws so as to force the jaws to closed position into clamping engagement with the article. It will be understood that two or more of the fasteners are employed for hanging sheets or other large articles upon the line. As the fastening devices may freely swing with respect to the line clamps, owing to the loose engagement of the terminal eyes 18 of the clamps, it will be manifest that the articles will not be blown from the fasteners by high winds but will readily sway or swing with respect to the line. . . .

UNITED STATES PATENT OFFICE

TAPEWORM-TRAP

Specification of Letters Patent No. 11,942, dated November 14, 1854

. . . The object of my invention is to effect the removal of worms from the system, without employing medicines, and thereby causing much injury.

My invention consists in a trap which is baited, attached to a string, and swallowed by the patient after a fast of suitable duration to make the worm hungry. The worm seizes the bait, and its head is caught in the trap, which is then withdrawn from the patient's stomach by the string which has been left hanging from the mouth, dragging after it the whole length of the worm. . . .

The trap is baited by taking off the cover b, of the exterior box, and filling the interior box with the bait which may consist of any nutritious substance. The interior box d, is then pushed down until the stud f, catches between the teeth of the opening e, and holds it with the openings, e, and c, opposite each other, the points of the teeth being then below the lower edge of the opening c. The trap, having the cord h, attached to a ring i, on the lid is then swallowed. The worm, in inserting its head at the opening e, and eating the bait, will so far disturb the inner box as to work it free of the stud f, when the box will be forced upward by the spring g, and the worm caught behind the head, between the serrated lower edge of the opening in the interior box, and the upper edge of the opening in the exterior box. The trap and the worm may then be drawn from the stomach, by the cord h. . . .

DEVICE FOR PREVENTING HENS FROM SETTING.

No. 582,320.   Patented May 11, 1897.

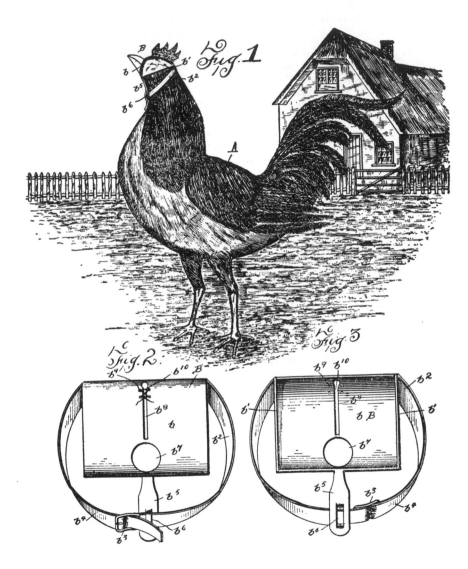

Hoods for Setting Hens

UNITED STATES PATENT OFFICE

DEVICE FOR PREVENTING HENS FROM SETTING

Specification forming part of Letters Patent No. 582,320, dated May 11, 1897
Application filed October 15, 1896. Serial No. 608,953. (No model)

. . . This invention relates to improvements in devices for preventing hens from setting; and it consists of certain novel constructions, combinations, and arrangements of parts, all of which will be hereinafter more fully set forth and claimed.

In the accompanying drawings, forming part of this specification, Figure 1 represents a side elevation of a hen with my invention applied thereto. Fig. 2 represents an enlarged front elevation of my device, and Fig. 3 represents an enlarged rear elevation of the same. . . .

When this device is to be applied to a hen, the hood is slipped over her head with the comb projecting through the comb-slot and the bill through the bill-aperture. The strip b^4 is then buckled through the buckle b^3 and the hood thus secured firmly in position.

When a hen is provided with one of these improved hoods, she can neither see to the right nor the left nor upward, and she is thus prevented from flying to any elevated position.

All nests in the modern construction of henneries are constructed at an elevation from the ground, and as a fowl will never fly in a direction in which it cannot first look the hen will thus be prevented from flying up into the nest.

The device will also prevent fowls from flying over fences and into gardens and the like. When a hen is provided with one of these hoods, the action of eating is not interfered with at all, as she can look downward and toward the ground as readily as she could were the hood not in position at all.

It will be observed that while the hen is prevented from reaching the nest or flying over fences she is perfectly free to scratch about or eat at pleasure. . . .

SIPHON SPOUT.

(Application filed Nov. 10, 1899.)

(No Model.)

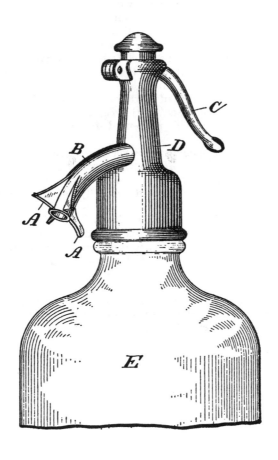

Hygienic Siphon

UNITED STATES PATENT OFFICE

SIPHON-SPOUT

Specification forming part of Letters Patent No. 641,201, dated January 9, 1900
Application filed November 10, 1899. Serial No. 736,565. (No model)

. . . My invention relates to projections on the mouth of the spout or nozzle of siphons; and the object of my improvement is to prevent the use of the siphon as a syringe by inserting it in or applying it to certain parts of the human body. I attain this object by the mechanism illustrated in the accompanying drawing, in which the diagram is a side view of the siphon. . . .

In using the siphon before my invention nothing prevented the insertion of the siphon-mouth in any opening of the body to be used as a syringe, thereby contaminating it with germs of disease. With my improvement in using the siphon as a siphon the projections *a* in no way interfere; but if it is attempted to misuse the siphon—*e. g.,* as a syringe—the projections *a* prevent an insertion in or application to any part of the body. . . .

Fig. 1

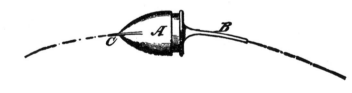

Fig. 2.

IMPROVEMENT IN PROJECTILES

Letters Patent No. 107,909, dated October 4, 1870
The Schedule referred to in these Letters Patent and making part of the same.

. . . My invention has for its object to furnish an improvement in balls and other projectiles, by means of which the ball or other projectile may be fired in curved lines with the same accuracy as in straight lines; and

It consists in constructing the ball with a curved flat piece upon its base, whether used with or without a curved flat point. . . .

ATTACHMENT FOR LOCOMOTIVES.

No. 292,504.   Patented Jan. 29, 1884.

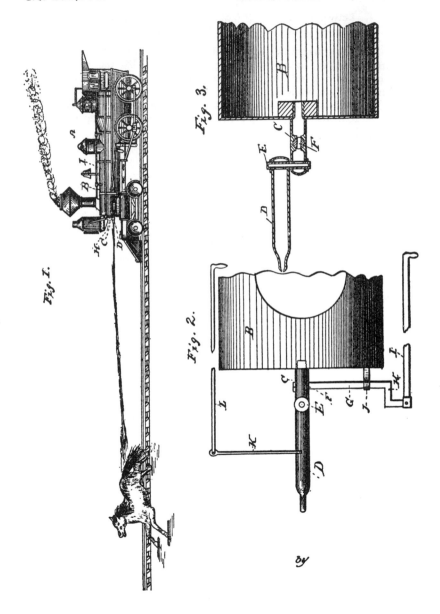

Fig. 1.

Fig. 2.

Fig. 3.

by

Device for Driving Cattle from Tracks

UNITED STATES PATENT OFFICE

ATTACHMENT FOR LOCOMOTIVES

Specification forming part of Letters Patent No. 292,504, dated January 29, 1884

Application filed March 26, 1883. (No model)

. . . This invention relates to an attachment for locomotives, to be used for frightening horses and cattle off the track. . . .

By means of the rod I the stop-cock F may be opened, thus permitting the water to escape from the boiler through the nozzle D, through which it is driven, by the steam-pressure in the boiler, with a great degree of force, and to a considerable distance, so that it may be employed for frightening horses and cattle off the track. By means of the rod L the nozzle may be adjusted so as to throw the stream of water in other than a straight line, so that the device may be advantageously used on curves. . . .

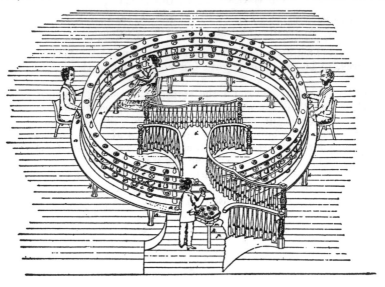

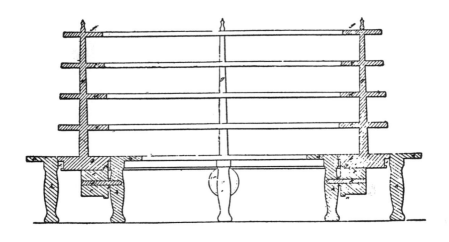

Self-Service Table

UNITED STATES PATENT OFFICE

IMPROVED SERVING-TABLE

Specification forming part of Letters Patent No. 55,677, dated June 19, 1866

. . . The nature of my improvement consists in providing a revolving or moving serving-table, *b*, driven by steam or other power, with either two or three rings or continuous tables, *a a*, for dining and other uses . . . the outer ring *a* and inner ring *a* fixed or stationary, having an open space, on the inside of the table *a*, for the seating of guests, as well as upon the outside of the table *a*, with an entrance or passage-way either above or below the tables *a a* and *b*, the middle or center ring *b* or continuous table *b*, with one or more shelves, *c d e f*, or a succession of shelves one above another, to revolve or move within or between the stationary rings or continuous tables *a a*, and so arranged as to seat persons conveniently at either the inner or outer rings or stationary tables *a a*, the central revolving or moving table *b*, with shelves *c d e f*, arranged one above another, loaded with the entire bill of fare once in every fifteen or twenty feet or sections of the table *b*, and the table *b*, with shelves *c d e f*, so loaded with viands, to move at the rate of fifteen or twenty feet per minute, or to pass before each guest at such speed as to exhibit before each guest the entire bill of fare once per minute. . . .

All persons at this table are put upon an equality and free to act for themselves, and these shelves so arranged as not only to contain the full bill of fare, and that kept hot by lamp or otherwise, but also to contain all the necessary dishes, knives, forks, spoons, glasses, &c., and also so arranged as to carry the dishes that have been used off into the pantry P, behind the screen, where they are removed by the servant stationed at that point for that purpose, and where also are the persons stationed to supply and replenish the revolving or moving table *b*, with shelves *c d e f*. The carver and his assistants are also stationed behind the screen, which we here term "pantry," P, to supply continually the revolving or moving table *b* and shelves *c d e f*. The dishes, after the guest has finished with them, are put upon the lower shelf or table *b*, which is hid from view by means of a lid or curtain.

All the servants that we require in the use of this moving-table is one upon the outside and one upon the inside, except those required in the pantry to put away the last dishes of each guest and brush off the crumbs and adjust the chair. This would be the requirements of a table that would seat, say, one hundred and fifty persons. . . .

DEVICE FOR WAKING PERSONS FROM SLEEP.

No. 256,265.   Patented Apr. 11, 1882.

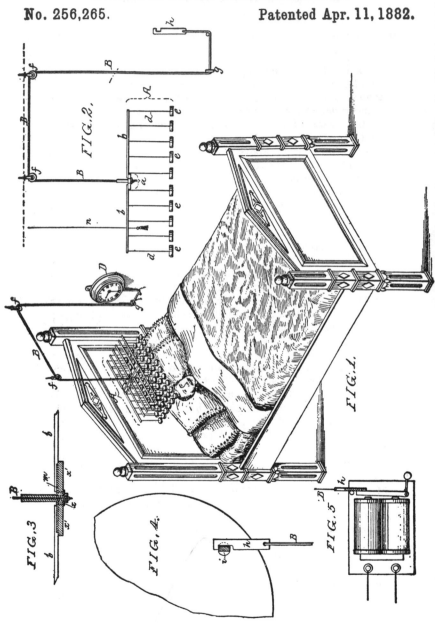

Tactile Alarm Clock

UNITED STATES PATENT OFFICE

DEVICE FOR WAKING PERSONS FROM SLEEP

Specification forming part of Letters Patent No. 256,265, dated April 11, 1882
Application filed December 14, 1881. (No model)

. . . The object of my invention is to construct a simple and effective device for waking persons from sleep at any time which may have previously been determined upon, the device being also adapted for use in connection with an electric or other burglar-alarm apparatus, in place of the usual gong-alarms. . . .

Ordinary bell or rattle alarms are not at all times effective for their intended purpose, as a person in time becomes so accustomed to the noise that sleep is not disturbed when the alarm is sounded.

The main aim of my invention is to provide a device which will not be liable to this objection.

In carrying out my invention I suspend a light frame in such a position that it will hang directly over the head of the sleeper, the suspending-cord being combined with automatic releasing devices, whereby the frame is at the proper time permitted to fall into the sleeper's face.

In the drawings, A represents the frame, which consists of a central bar, *a*, having on each side a number of projecting arms, *b*, the whole being made as light as is consistent with proper strength. From each of the arms *b* hang a number of cords, *d*, and to the lower end of each of these cords is secured a small block, *e*, of light wood, preferably cork . . . the only necessity to be observed in constructing the frame being that when it falls it will strike a light blow, sufficient to awaken the sleeper, but not heavy enough to cause pain. . . .

DEVICE FOR PRODUCING DIMPLES.

No. 560,351.   Patented May 19, 1896.

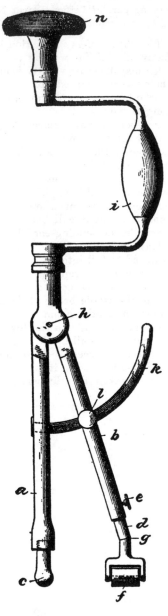

Dimple Tool

UNITED STATES PATENT OFFICE

DEVICE FOR PRODUCING DIMPLES

Specification forming part of Letters Patent No. 560,351, dated May 19, 1896

Application filed May 25, 1895. Serial No. 550,658. (No model)

. . . The present invention consists of a device which serves either to produce dimples on the human body or to nurture and maintain dimples already existing.

In order to make the body susceptible to the production of artistic dimples, it is necessary, as has been proved by numerous experiments, that the cellular tissues surrounding the spot where the dimple is to be produced should be made susceptible to its production by means of massage. This condition is fulfilled by the present process as well as by the apparatus by which the process is worked, and which is represented in an enlarged form in the accompanying drawing. . . .

When it is desired to use the device for the production of dimples, the knob or pearl c of the arm a must be set on the selected spot on the body, the extension d, together with the cylinder f, put in position, then while holding the knob n with one hand the brace i must be made to revolve on the axis x. The cylinder f serves to mass and make the skin surrounding the spot where the dimple is to be produced malleable. . . .

ANIMAL TRAP.

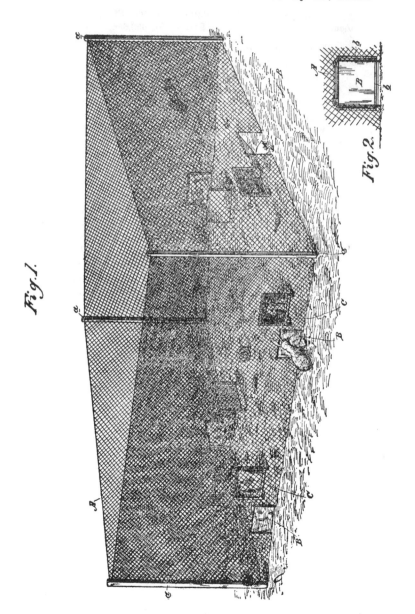

Fig.1.

Fig.2.

UNITED STATES PATENT OFFICE

ANIMAL-TRAP

Specification forming part of Letters Patent No. 383,700, dated May 29, 1888
Application filed February 16, 1888. Serial No. 264,231. (No model)

. . . The object of my invention is to provide an effective trap for rabbits, and one which is at the same time cheap in construction and readily transportable. . . .

Any suitable enticing material or food may be placed within the cage. The rabbit, endeavoring to gain admission to the cage, soon sees what he deems an unguarded opening, the door B being of glass, and therefore not noticed by him. He therefore makes for the opening, and the door, swinging inwardly, does not impede him, and under his original impulse he passes through into the cage. He cannot get out again, for the doors B do not swing outwardly.

Now, in order to better insure the entrance of the rabbit, I place within the cage and just behind each door B a mirror, C, the location being such that upon discovering and approaching the supposed opening of the door B the rabbit cannot fail to observe his image in the mirror. Surprised by this he pricks up his ears, and, his image doing likewise, he is the more impelled to enter the cage, in order to make the acquaintance of so close a companion, for though other rabbits might really be within and in sight, still the proximity and sympathetic actions of the reflected rabbit do more to create a sudden impulse toward the door B than the confined real rabbits do. . . .

Neck-tie & Watch Guard.

Nº 79063.   Patented Jun. 23. 1868

Combination Cravat and Watch-Guard

UNITED STATES PATENT OFFICE

IMPROVED NECK-TIE AND WATCH-GUARD COMBINED

Letters Patent No. 79,063, dated June 23, 1868

The Schedule referred to in these Letters Patent and making part of the same.

. . . My invention has for its object to combine a neck-tie and watch-guard with each other, so as to furnish a neat, convenient, and serviceable article; and it consists in combining a neck-tie and watch-guard in one article, as hereinafter more fully described. . . .

In using the article, the middle part, a^1, is passed around the neck of the wearer, and the knot or bow a^3 slipped up to its place. The elastic loop a^4 is then passed over the front button of the shirt-neck band, to keep the said knot from slipping down out of place. The guard-ring of the watch may then be attached to the extreme ends of the neck-tie A by a snap-ring or other means, or the watch may be permanently attached to said ends. . . .

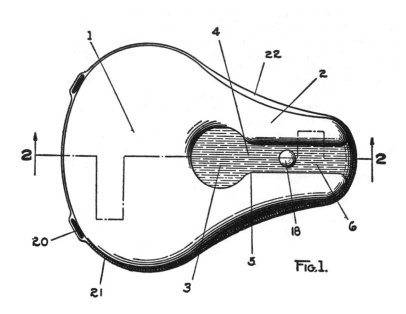

Fig.1.

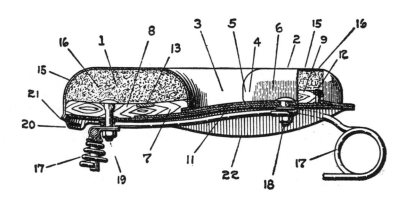

Fig.2.

Bicycle Saddle for the Use of Either Sex

UNITED STATES PATENT OFFICE

BICYCLE AND MOTOR-CYCLE SEAT

Patented May 19, 1925   1,538,542

Application filed February 16, 1924. Serial No. 693,327

. . . It is a primary object of my invention to provide a bicycle or motorcycle saddle having a suitable cavity properly located to allow comfortable clearance for the private organs of the male rider, said saddle having also a channel adapted to allow clearance for the female rider's private organs, to prevent pressure at the opening of said organs due to the weight of the rider, and tending also to keep said organs in a naturally closed state, the sides of said channel being substantially parallel and bell-mouthed. . . .

Referring to Fig. 1, a somewhat circular cavity 3 is formed on a medial line of the apparatus as a whole, at the junctions of said extension with said larger seating portion. Said cavity is of suitable size, shape and location to comfortably receive the private organs of a male rider and more particularly the testicle region of such rider. Said opening is of bell-mouthed formation, the bell-mouth character thereof being formed on the upper portion of said opening. Such bell-mouthed formation is particularly useful to the comfort of the male rider both during the riding act and also during the mounting or dismounting acts, said organs being slidably lodgeable or dislodgeable in relation to said opening when same is thus covered or uncovered by angular movement as compared to what may be called a vertical straight-away movement. . . .

1,278,217.

Patented Sept. 10, 1918.

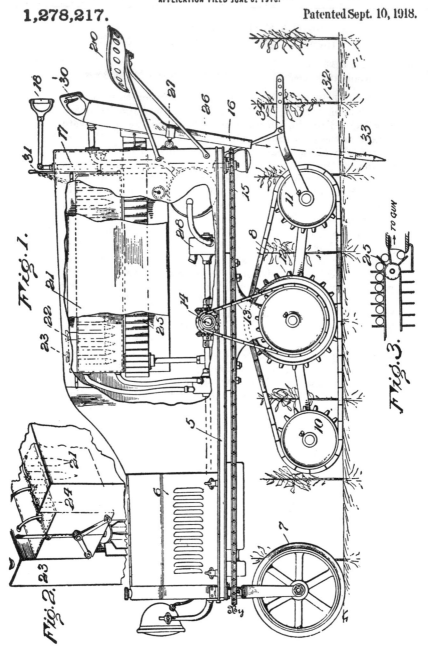

UNITED STATES PATENT OFFICE

APPARATUS FOR IRRIGATING PLANTS

l,278,217   Specification of Letters Patent   Patented Sept. 10, 1918
Application filed June 8, 1918. Serial No. 238,863

. . . This invention relates to irrigating apparatus and more particularly to an apparatus especially adapted for irrigating growing trees and plants of all kinds in climates where surface irrigation, due to extremely high temperatures and dry atmospheres, is impracticable for obtaining the best results. . . .

In operation, the machine travels along between or straddling the rows of plants under its own power and as the plants are reached the gun 26 is fired to discharge a projectile of ice into the ground at the roots of the plant, as shown. The apparatus is of such size as to provide twenty or thirty projectiles at once to the endless belt 25 and while these are being used, a second charge will have been made by the refrigerant in the freezing tank. When not used for the above purpose the apparatus may be used wherever a tractor is applicable. . . .

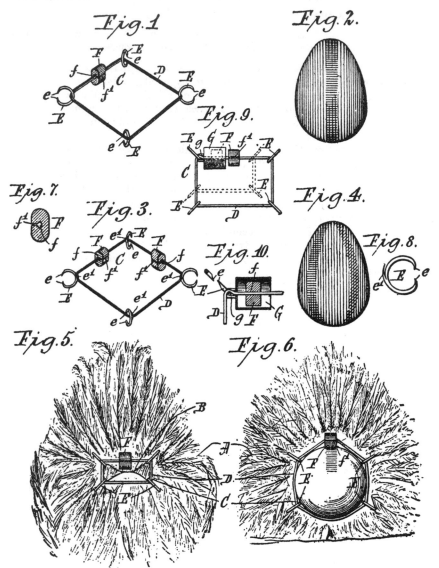

Fig.1.

Fig.2.

Fig.9.

Fig.7.

Fig.3.

Fig.10.

Fig.4.

Fig.8.

Fig.5.

Fig.6.

UNITED STATES PATENT OFFICE

EGG-MARKING DEVICE

970,074   Specification of Letters Patent   Patented Sept. 13, 1910
            Application filed April 18, 1910. Serial No. 556,049

. . . The primary object of my invention is the production of an egg marking-device bearing a marking-element or elements whereby the eggs laid by the hen to which the marking device is attached will be marked in a distinctive manner and whereby the laying capacities or qualities of each hen in a hennery can be easily ascertained.

Another object of my invention is the provision of a marking-device of this character which can be easily attached to the vent of a hen so that it will always be in place for marking an egg laid by said hen.

A further object of my invention is to so construct the marking-device that it will yield with the walls of the vent as the egg is being laid, thus permitting the egg to pass through the marking-device. . . .

When the egg is passed from the vent and through the supporting-band, said band is expanded, causing the marking element to be drawn out of the shield or protector, as best shown in Fig. 9, and as soon as the egg is laid, the walls of the vent return to normal position, as shown in Fig. 5, so that the supporting-band is relieved of tension and the marking-element enters the shield or protector. . . .

SALUTING DEVICE.

No. 556,248. Patented Mar. 10, 1896.

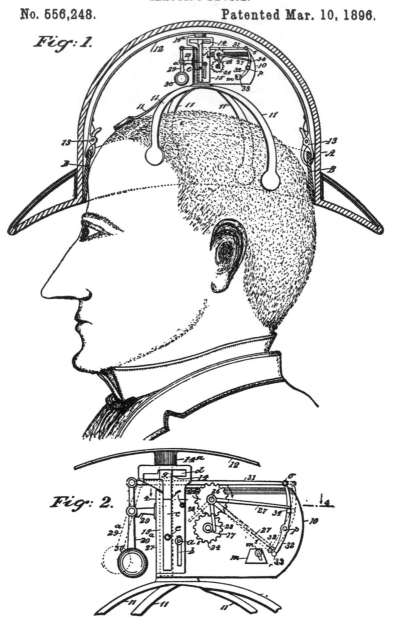

Fig: 1.

Fig: 2.

Hat-Tipping Device

UNITED STATES PATENT OFFICE

SALUTING DEVICE

Specification forming part of Letters Patent No. 556,248, dated March 10, 1896
Application filed September 18, 1895. Serial No. 562,908. (No model)

... This invention relates to a novel device for automatically effecting polite salutations by the elevation and rotation of the hat on the head of the saluting party when said person bows to the person or persons saluted, the actuation of the hat being produced by mechanism therein and without the use of the hands in any manner. . . .

Should the wearer of the hat having the novel mechanism within it and engaging his head, as before explained, desire to salute another party, it will only be necessary for him to bow his head to cause the weight-block 30 to swing forwardly. The swinging of the block 30, as stated, will, by the consequent vibration rearwardly of the upper end of the arm 29^a, push the rod 31 backward and release the stud 34 on the rock-arm 32 from an engagement with the lifting-arm 27, so that the latter will, by stress of the spring 16, be forcibly rocked down into contact with the pin 33, as indicated by dotted lines in Fig. 2, the arm 28 having been correspondingly moved toward the lift-pin f, as also shown by dotted lines in the same figure. When the person making a salutation with the improvement applied to his hat resumes an erect posture after bowing, the weight 30 will swing back into a normal position, which will draw the upper end of the rock-arm 32 forwardly and move its lower end rearwardly far enough to release the arm 27 from the pin 33. The gear-wheel 25 will now be moved by the spring 16, so as to impinge the short arm 28 on the lower side of the stud f, which will cause the guide-plate 15 to slide upward, carrying the post 14 with it. Just before the arm 28 passes the stud f the detent-spring q will press its curved toe q' through the slot in the front plate of the case 10 and project said toe below the rounded lower end of the post 14. The lifting-arm 27 is now brought into contact with the pin e, and the pressure of the said arm on the pin e causes the post 14 to move upwardly in the depression c of the guide-plate 15 until it enters the slot d. The lift-pin e will now be swung through the rear portion of the cross-slot d by the arm 27, and by the impetus given to the pin and post 14 by said arm the post, bow-piece, and hat A will receive a rotary movement sufficient to bring the pin e into the depression c, when the gravity of the parts will cause the hat to drop into its normal position on the wearer's head. . . .

EMERGENCY RUBBER OVERSHOE

Filed June 8, 1923

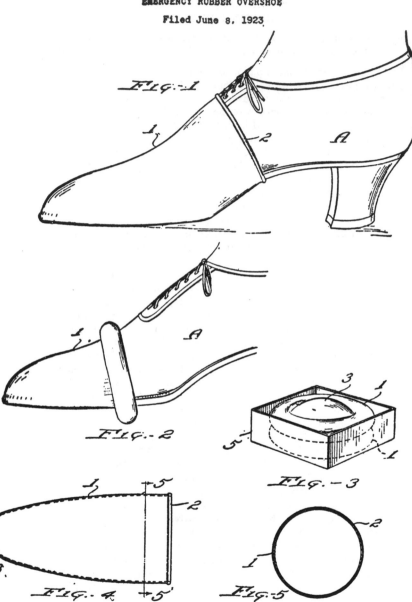

FIG.-1

FIG.-2

FIG.-3

FIG.-4

FIG.-5

Emergency Rubber Overshoe

UNITED STATES PATENT OFFICE

EMERGENCY RUBBER OVERSHOE

Patented Oct. 19, 1926   1,603,923

Application filed June 8, 1923. Serial No. 644,074

. . . This invention relates to a novel form of overshoe and has for its essential objects the provision of a very effective emergency rubber overshoe, adapted to be uniformly rolled up into a very small space and which may be so cheaply manufactured and sold as to invite the purchase and wearing of a pair on a single occasion, but which shall, at the same time, be sufficiently durable to be worn on numerous occasions if desired.

A further and important object is to so construct such an overshoe that when rolled into its compact space, it is ready for most immediate and easy application to the shoe and may be correspondingly easily removed by a rolling movement, bringing it again to the small compact form ready for subsequent wear.

A further object is to make the overshoe of very light construction using a very small amount of resilient material, thereby accomplishing economy and in addition causing the overshoe, by reason of its thinness and resiliency, to fit neatly in any position. To this end, I make the overshoe of substantially uniform thickness throughout, whereby it may be rolled on and will invariably position itself to neatly fit the shoe, and no regard need be given to the idea of rights or lefts, or even to the thought of which is top or bottom. Further advantageous and unique characteristics of my overshoe will appear in the following description, which relates to the accompanying drawings. The essential characteristics are summarized in the claims. . . .

When the overshoes are rolled up as indicated at Fig. 3, they take the form of a rolled ring with the portion 3 across the center thereof, and may be nested very neatly in a small package or box, such, for example, as indicated at 5. From this position they may be placed upon the shoes with great facility as follows:—

The ring is placed over the toe with the dome shaped portion 3 fitting over the convexity of the toe, and by a rolling motion between the thumb and finger, drawing rearwardly on the shoe, the rubber wall of the shoe is drawn tightly and evenly into position, until the bead 2 comes to the position shown in Fig 1, the operation requiring but a brief moment. . . .

ELECTRICAL BEDBUG EXTERMINATOR.

(Application filed Feb. 7, 1898.)

(No Model.)

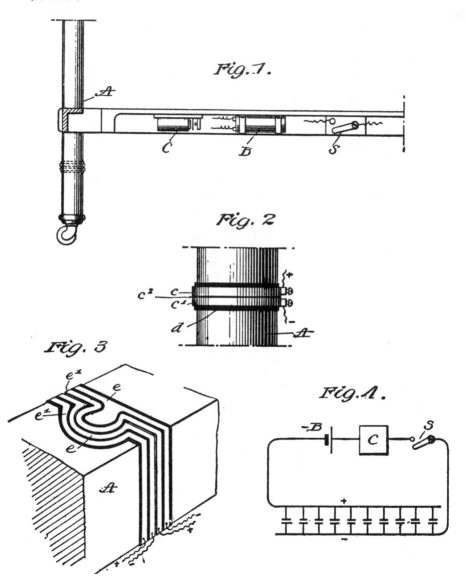

Fig. 1.

Fig. 2

Fig. 3

Fig. 4.

Vermin Electrocutor

UNITED STATES PATENT OFFICE

ELECTRICAL BEDBUG-EXTERMINATOR

Specification forming part of Letters Patent No. 616,049, dated December 13, 1898
Application filed February 7, 1898. Serial No. 669,334. (No model)

. . . This invention relates to bedbug-exterminators; and it consists of electrical devices applied to bedsteads in such a manner that currents of electricity will be sent through the bodies of the bugs, which will either kill them or startle them, so that they will leave the bedstead. The electrical devices used consist of a battery, induction-coil, a switch, and a number of circuits leading to various locations on the bedstead, where are placed suitable circuit-terminals, arranged so that the bugs in moving about will close the circuit through their own bodies. . . .

The space c^2 between the rings is such that a bug in crossing from one to the other must close the circuit through its own body, and thus receive a current of electricity. If these rings are placed on a leg of the bedstead, an insect in climbing up will when it receives the shock more than likely change its mind and return in the direction whence it came. Another location where the contacts would be particularly efficient is at the joints between the side pieces and the head and foot boards. A perspective of such a joint is shown in Fig. 3. A pair of insulated contact-strips e and e' is placed along each of the contiguous edges of the joint and surrounding the joint on all sides. The polarities of these strips are so arranged that a positive and negative strip will be next to the respective edges, so that the insect in crossing a pair or the adjacent members of the two pairs will necessarily receive a current, which will either terminate its career at once or make it seek other locations. In like manner contact-strips in pairs, constituting the terminals of the circuit, may be located at various places on the bedstead or on the bedsprings, which will so harass the bugs as to cause them to shun the bed entirely. . . .

ALARM

Filed Dec. 6, 1927

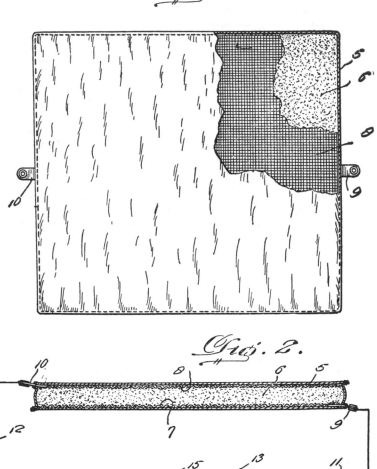

Bed-Wetting Alarm

UNITED STATES PATENT OFFICE

ALARM

Patented Aug. 5, 1930   1,772,232

Application filed December 6, 1927. Serial No. 238,028

. . . This invention relates to alarms, and has more particular reference to an electrical device of this kind adapted for use as an aid in curing persons of the habit of nocturnal urination or bed-wetting.

The primary object of the invention is to provide means for sounding an alarm when urination starts so as to awaken the sleeper before the bladder is emptied and thereby enable him or her to avoid material wetting of the bed. . . .

In operation, the filling material is normally dry and insulates the sheets 7 and 8 from each other so as to prevent sounding of the bell 17 even though the switch 15 is closed. When the sleeper begins to urinate, the urine runs through the casing 5 and conductor sheet 8 and saturates a portion of the filling material 6 entirely through to the sheet 7. The wetted portion of the material 6 is thus rendered conductive to electrically connect the sheets 7 and 8 and close the alarm circuit, whereby the bell 17 is caused to operate and awaken the sleeper before his or her bladder is emptied. By turning the switch 15 off, the alarm may be rendered silent, and by constructing the casing so that it may be opened, the wet filling material may be replaced by dry material to condition the device for re-use. . . .

COMBINED MATCH SAFE, PINCUSHION, AND TRAP.

No. 439,467. Patented Oct. 28, 1890.

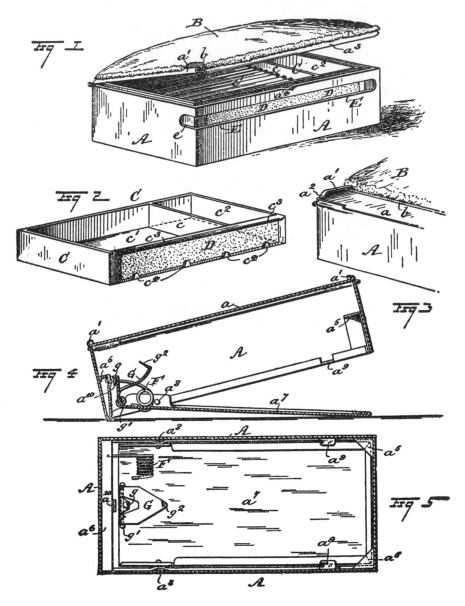

Triple Utility Device

UNITED STATES PATENT OFFICE

COMBINED MATCH-SAFE, PINCUSHION, AND TRAP

Specification forming part of Letters Patent No. 439,467, dated October 28, 1890
Application filed May 27, 1890. Serial No. 353,345. (No model)

. . . My invention relates to a device adapted for a match-safe, pincushion, and trap, and has for its object to provide a simple, inexpensive, and compact structure capable of effective use for any of the purposes above specified.
. . .

When the device is to be set as a trap, the top pincushion B will be slid forward off the lid a of the main box or casing A and the match-tray C will be removed from the casing, and after the box-bottom is opened downward against the tension or torsional strain of the spring F and the lip g of the baited hook is set lightly upon the casing-catch a^{10} to hold the bottom open the casing, with its top a latched or closed, will be set on the floor or on a shelf, as shown in Fig. 4 of the drawings. When a mouse walks up the bottom a^7 and nibbles the bait on the hook G, the hook-lip g will be tripped from its detent a^{10}, and the spring F then will instantly close the bottom as the box or casing falls flat, and the mouse or animal will be caught in the trap. As the pincushion B is removed from the casing-top the casing, with the mouse in it, may be immersed in water to drown the animal without damaging the cushion, and after use of the device as a trap the match-tray C may again be put into the casing, and the cushion B will be again slipped into the guides a' on the casing top or lid, and the device is again ready for its most ordinary uses as a match-safe and pincushion, as will readily be understood.

91

METHOD OF PRESERVING THE DEAD.

APPLICATION FILED OCT. 13, 1903.

NO MODEL.

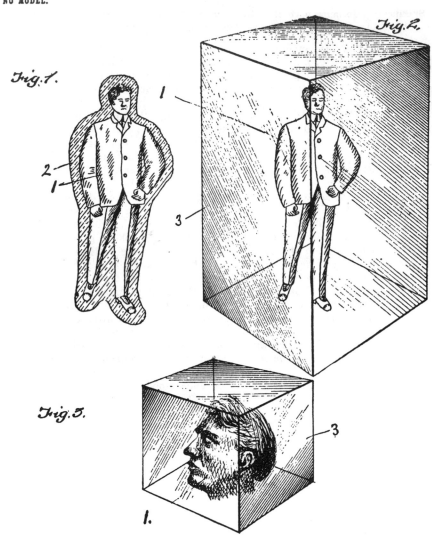

Fig.1.

Fig.2.

Fig.3.

UNITED STATES PATENT OFFICE

METHOD OF PRESERVING THE DEAD

No. 748,284   Patented December 29, 1903

Specification forming part of Letters Patent No. 748,284, dated December 29, 1903
Application filed October 13, 1903. Serial No. 176,922. (No specimens)

. . . This invention relates to certain new and useful improvements in methods of preserving the dead; and it has for its object the provision of a means whereby a corpse may be hermetically incased within a block of transparent glass, whereby being effectually excluded from the air the corpse will be maintained for an indefinite period in a perfect and life-like condition, so that it will be prevented from decay and will at all times present a life-like appearance. . . .

In carrying out my process I first surround the corpse 1 with a thick layer 2 of sodium silicate or water-glass. After the corpse has been thus inclosed within the layer of water-glass it is allowed to remain for a short time within a compartment or chamber having a dry heated temperature, which will serve to evaporate the water from this incasing layer, after which molten glass is applied to the desired thickness. This outer layer of glass may be molded into a rectangular form 3, as shown in Fig. 2 of the drawings, or, if preferred, cylindrical or other forms may be substituted for the rectangular block which I have illustrated. In Fig. 3 I have shown the head only of the corpse as incased within the transparent block of glass, it being at once evident that the head alone may be preserved in this manner, if preferred. . . .

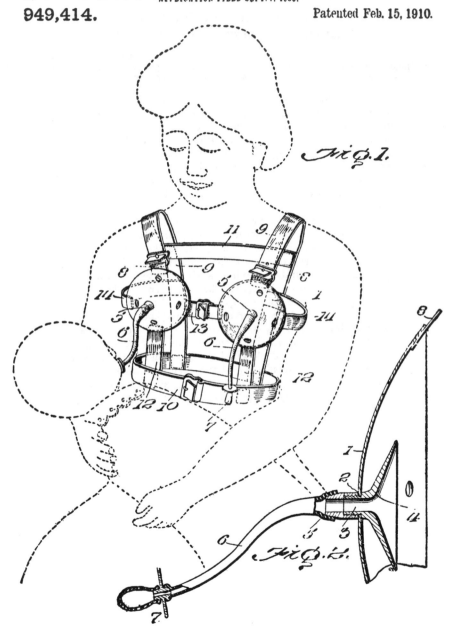

Fig. 1.

Fig. 2.

Device for Modest Mothers

UNITED STATES PATENT OFFICE

NURSING ATTACHMENT

949,414   Specification of Letters Patent   Patented Feb. 15, 1910

Application filed September 7, 1909. Serial No. 516,336

. . . The primary object of this invention is an improved construction of device for use by mothers with nursing infants, and designed particularly to avoid unpleasant and embarrassing situations in which mothers are sometimes placed in public places by the necessary exposure of the breast in suckling the child.

With this and other objects in view, as will more fully appear as the description proceeds, the invention consists essentially in a nursing attachment designed to be worn over the breasts and arranged for the detachable connection thereto of the nipple on a tube of any desired length, the nipple or nipples, according to whether there be one or two employed, being worn inside of the shirtwaist or other outer garment and it being only necessary when the child is to be nursed, to slip the nipple out of the waist, thereby avoiding the necessity of exposing the person. . . .

In the practical use of the device, the shields 1 are adjusted over the breasts with the cups 4 directly over the nipples and the cap or caps 5 are then attached, the nursing nipples 7 and all other parts being hidden beneath the wearer's waist. Whenever the child requires nursing it is only necessary to slide one of the nursing nipples 7 out from the waist and the child can obtain its proper nourishment without the exposure of the mother's person and the consequent embarrassment which is thus often occasioned. . . .

1,051,684.

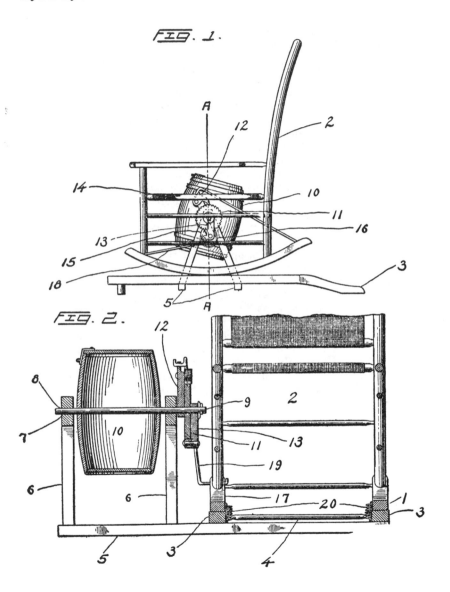

FIG. 1.

FIG. 2.

UNITED STATES PATENT OFFICE

CHURN

1,051,684   Specification of Letters Patent   Patented Jan. 28, 1913
Application filed June 28, 1912. Serial No. 706,431

. . . This invention relates to improvements in churns and especially with reference to improvements in means, actuated by a rocking chair to operate a churn, so that a churn may be operated by a person seated and rocking in a rocking chair, the invention consisting in the construction, combination and arrangement of devices, hereinafter described and claimed. . . .

When the chair is rocked, the rods 18—19 cause the arms 12—13 to be oscillated and to move simultaneously in the same direction, so that while the pawl 14 moves rearwardly on the upper side of the ratchet wheel and turns the latter the pawl 15 will slip on the under side of the ratchet wheel, and in the reverse movement of the said arms 12—13, the pawl 15 by engagement with the lower side of the ratchet wheel will turn the latter while the pawl 14 slips on the upper side thereof. Hence, the ratchet wheel is rotated in one direction and causes the churn body to correspondingly rotate. . . .

1,016,164.

Patented Jan. 30, 1912.

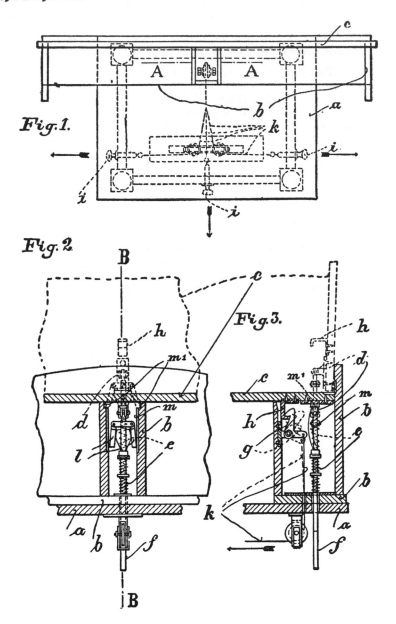

Jury Protector

UNITED STATES PATENT OFFICE

DEVICE FOR PROTECTING THE MEMBERS OF A COURT ASSEMBLY

1,016,164   Specification of Letters Patent   Patented Jan. 30, 1912
Application filed March 11, 1910. Serial No. 548,755

. . . The present invention relates to a device for the protection of the members of a court assembly against attempts upon their lives, such attempts being of frequent occurrence at the giving of verdicts.

The invention consists in the provision of a bullet-proof plate which is hingedly connected to the head piece of the court table and which, while being normally maintained in horizontal position, is actuated by a spring so as to be raised into vertical, protecting position as soon as released, such releasing being effected by the actuating knobs arranged in various places around the table. . . .

To the back side of the head piece b of a court table a, a bullet-proof plate c is hingedly connected. A bracket d on the underside of said plate is pivotally connected to a rod f which is slidably guided in the head piece b and encircled by a helical spring e, the latter abutting against a collar on the rod and tending to raise the rod for bringing the plate c into vertical position, as shown dotted in Fig. 3. The plate c is normally maintained in horizontal position by means of a pivoted catch g engaging a bent finger h on the underside of the plate. The catch g is of bell-crank-shape and can be actuated by means of a cord k so as to release the plate. This cord is guided by means of sheaves under the table top and branched off to the various seats of the table where the branches are connected to actuating knobs i. It is thus possible for any of the persons seated around the table to release the protecting plate which, as soon as released, assumes a vertical, protecting position. The rod f is also provided with pivoted spring-actuated detents l the ends of which slide in grooves m in vertical members of the head piece b and snap into recesses m^1 so as to secure the plate c in raised position. . . .

99

SANITARY COW STALL

Filed March 22 , 1924

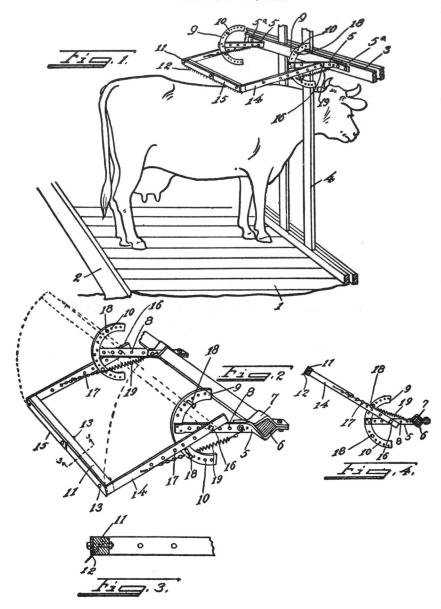

Device for Inhibiting Certain Acts in Cows

UNITED STATES PATENT OFFICE

SANITARY COW STALL

Patented Oct. 21, 1924   1,512,610

Application filed March 22, 1924. Serial No. 701,040

. . . My invention relates to cow stalls in which provision is made for preserving them in a clean condition.

It is a known fact that where a pointed object is arranged closely above the back of a cow, the animal will not be able to hump up while in the stall until she backs up far enough at the time of humping. As a result stalls may be kept clean from droppings.

The above known fact has been utilized in several instances as a basis for devices to be used in dairy barns where cleanliness of the animal, and particularly her udders, is a matter of vital importance. The object of my invention is to improve upon, and render more practical devices for such purpose. . . .

In use the operator will set the pivot bolts and the stop pins at such a point that the frame will lie closely above the proper point on the cow's back to result in pronging the cow with the pins of the pin bar, should she hump up, without backing to a position where the stall will not be soiled. . . .

When taking the cow out of the stall the dairyman merely throws up the frame, past center, whereupon it will spring up well beyond any chance contact with the animal, and when the cow is back in her place, he will drop the frame, so as to bring the pins into position, ready to interfere with the cow soiling her stall, by inhibiting the act of humping up. . . .

FEEDING DEVICE FOR POULTRY.
APPLICATION FILED MAR. 21. 1906.

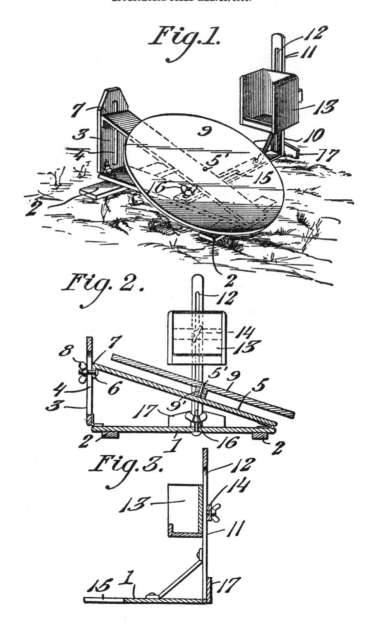

Fig.1.

Fig. 2.

Fig.3.

UNITED STATES PATENT OFFICE

FEEDING DEVICE FOR POULTRY

No. 828,227   Specification of Letters Patent   Patented Aug. 7, 1906

Application filed March 21, 1906. Serial No. 307,270

. . . The invention relates to an improvement in feeding devices for poultry, comprehending specifically an exerciser and feeder in the use of which the fowls secure a beneficial amount of exercise in feeding.

The main object of the present invention is the production of a device so constructed and arranged that the fowl in order to secure the food carried by the stretcher is compelled to undergo increased exercise as compared with the ordinary manner of feeding, whereby the fowls in the act of feeding are given that degree of exercise best suited to insure their proper condition. . . .

In operation the fowls desiring the food in the box and stepping upon the platform cause the same to revolve under their weight, with the result that they are compelled to move rapidly forward to maintain such position on the platform as will enable them to reach the food in the box. It is of course to be understood that in all relative positions of the parts the supporting-plate 5, and therefore the platform, is disposed at an angle to the base 1, whereby to provide for the necessary movement of the platform under the weight of the fowl. The angular adjustment of the plate 5, however, will permit of the angular arrangment of the supporting-plate being adjusted to accommodate the device to the weight of the particular species of fowls, it being of course evident that the heavier the fowl the less the relative inclination of the platform, as it is only desired that the weight of the fowl cause a revolution of the platform during the attempting of the fowl to reach the feed-box. . . .

1,180,753.

Patented Apr. 25 1916.

FIG.1.

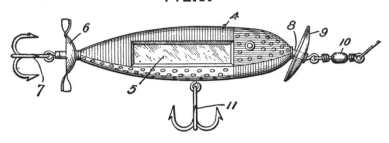

FIG.2.

UNITED STATES PATENT OFFICE

ARTIFICIAL FISH-BAIT

1,180,753   Specification of Letters Patent   Patented Apr. 25, 1916

Application filed April 23, 1915. Serial No. 23,351

. . . My invention relates more particularly to artificial baits for trawling, and its primary objects are to make the bait more attractive to the fish, to secure its proper position in the water, to provide a convenient and effective hanging of the hooks, and to generally improve the structure and operation of trawling baits. . . .

The attractiveness of the bait for the fish is increased by making the disks 6 and 9 of highly polished metal. The glitter and flashing lights occasioned by these and by the mirror are well known attractives; but the mirror 5 is an additional feature that insures the effectiveness of the bait in the following manner: A male fish seeing his image upon looking therein will appear to see another fish approach it from the opposite side with the intent to seize the bait, and this will not only arouse his warlike spirit, but also appeal to his greed, and he will seize the bait quickly in order to defeat the approaching rival. In case the fish is suspected of cowardice I may make the mirror of convex form, as shown at 5ª, in order that the rival or antagonist may appear to be smaller. In the case of a female fish the attractiveness of a mirror is too well known to need discussion. Thus the bait appeals to the ruling passion of both sexes, and renders it very certain and efficient in operation. . . .

FISHING APPARATUS.

No. 515,001. Patented Feb. 20, 1894.

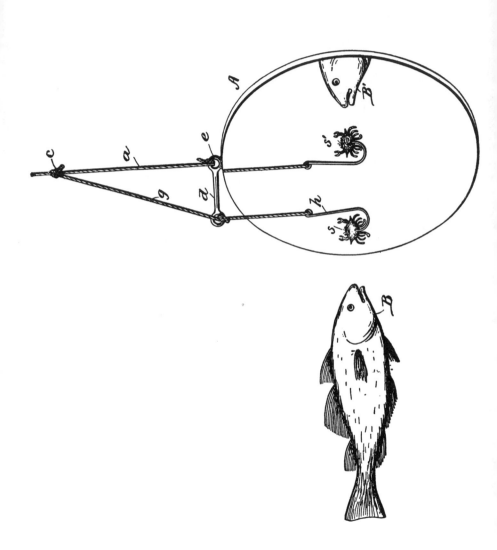

Double Bait

UNITED STATES PATENT OFFICE

FISHING APPARATUS

Specification forming part of Letters Patent No. 515,001, dated February 20, 1894
Application filed April 29, 1893. Serial No. 472,314. (No model)

. . . This invention relates to that class of devices used as decoys in fishing, the object of it being to induce the fish to take the bait more readily. It is illustrated in the accompanying drawing, which may be explained as follows:

. . . In using the apparatus, a bait s, (represented in this case as a small crab) is put on the hook h, and let down into the water with the mirror which serves as a sinker, until its lower edge just touches the bottom. In this position, the least pull on the hook on the branch line, will be felt very plainly by the hand at the upper end of the taut main line a. In this position, as shown in the drawing, the fish B, when approaching the bait s, will see the reflection B', of himself in the mirror, also coming for the reflection of the bait s', and will be made bolder by the supposed companionship, and more eager to take the bait before his competitor seizes it. He will lose his caution, and take the bait with a recklessness that greatly increases the chances of his being caught on the hook. The reflection of light from the mirror in the water, will have in some degree the effect that the lighted torch has in some well known kinds of fishing, of attracting fish to the bait, and the light reflected by the mirror upon the bait, will make it more conspicuous. . . .

UNITED STATES PATENT OFFICE

IMPROVEMENT IN FIRE-ESCAPES

Specification forming part of Letters Patent No. 221,855, dated November 18, 1879
Application filed March 26, 1879

. . . This invention relates to an improved fire-escape or safety device, by which a person may safely jump out of the window of a burning building from any height, and land, without injury and without the least damage, on the ground; and it consists of a parachute attached, in suitable manner, to the upper part of the body, in combination with overshoes having elastic bottom pads of suitable thickness to take up the concussion with the ground. . . .

Fig. 1.

Fig. 2.

Fig. 3.

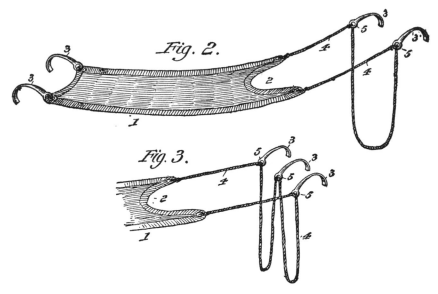

Day-Coach Sleeper

UNITED STATES PATENT OFFICE

HAMMOCK

Specification forming part of Letters Patent No. 400,131, dated March 26, 1889
Application filed June 19, 1888. Serial No. 277,573. (No model)

. . . This invention relates to an improvement in hammocks, and has special reference to a hammock so constructed that it can be slung from and used with the backs of the seats of ordinary railway passenger-cars.

The invention has for its object to provide a means whereby passengers who are obliged to travel in ordinary passenger-cars at night may be able to sleep with ease and comfort. . . .

Instead of employing the hammock as shown in Fig. 1, the occupant may recline practically at full length, by attaching the hooks on the rope 4 to the upper edge of the back of the adjacent seat after the back has been turned over. In this case the rope 4 will have been drawn through the eyes 3, thereby drawing up the loop, and the feet and lower portion of the legs of the occupant of the hammock will rest on the car-seat. . . .

1,033,788.

Patented July 30, 1912.

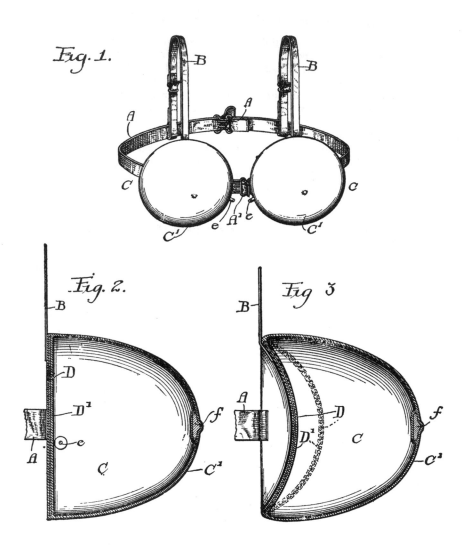

Fig. 1.

Fig. 2.

Fig 3

UNITED STATES PATENT OFFICE

BUST-FORM

1,033,788   Specification of Letters Patent   Patented July 30, 1912
Application filed August 6, 1908. Serial No. 447,194

. . . The invention relates to bust forms and designs to provide an improved device, designed to supply any natural deficiency of form and to support the breasts, which will effectively give the appearance of the natural bust, and which may be readily fitted to different breasts and worn with comfort. . . .

The invention further designs to construct the bust-forms so that when worn, they will not have the appearance of lifeless members, but will vibrate responsively to movements of the wearer. . . .

In use, when pneumatic pads are inflated and covered by garments, the bosom is sometimes rather lifeless in appearance, due to the inflated pads which do not readily or freely respond to the movements of the wearer's bosom. To overcome this appearance, a weight f is molded into the front of the casing C, and under the influence of the body is suspended at the front of the pad and sets the front in motion so that it will vibrate freely in response to any movement of the body of the wearer. . . .

No. 186,962.

Patented Feb. 6, 1877.

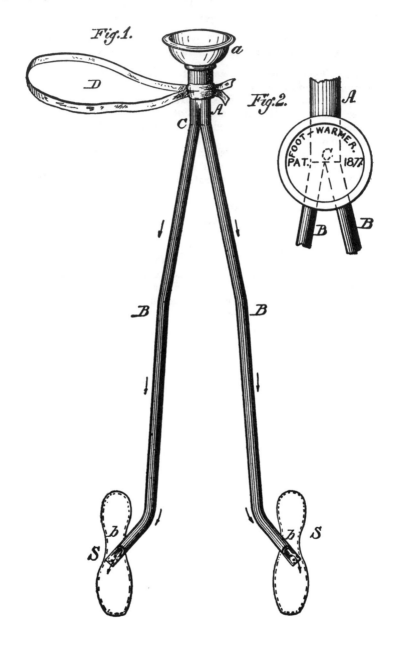

Pedal Calorificator

UNITED STATES PATENT OFFICE

IMPROVEMENT IN FOOT-WARMERS

Specification forming part of Letters Patent No. 186,962, dated February 6, 1877
Application filed January 16, 1877

. . . Be it known that I have made a new and useful invention in a "Pedal Calorificator" or "Foot-Warmer." . . .

It is a well-established fact that our lungs constitute the laboratory of nature, within which—by a condensing process—animal heat is generated, and afterward conveyed and distributed to other portions of our bodies by the action of the heart and circulation of the blood; that, for mechanical reasons the supply to the extremities—the hands and feet—on account of their distance from the center of heat, is more or less deficient, and, consequently, they suffer most when exposed to severe outward cold; the feet, especially, by reason of their immediate contact, in winter weather, with cold floors, as in railroad-cars and other vehicles, and with the frozen ground and icy sidewalks.

Now, I find, by personal experiment, that by breathing for a short time on the bulb of a thermometer I am enabled to raise the mercury to 88° Fahrenheit—only 10° below blood-heat—which I, therefore, assume to be the natural temperature of the breath, and . . . which, in the action of breathing, is totally dissipated and lost in the open air.

My invention aims at economizing and utilizing this wasted heat by any simple contrivance for conveying it to our feet, where it is so much needed. . . .

I have found, by actual experience, that the tubes in a short time become warmed by the body, so that little heat of the breath is lost in its passage to the feet; that, accordingly, the air I find is delivered in boot or shoe with a temperature of about 84° Fahrenheit—a loss of only 4°.

After a few sharp blasts of breath at the beginning—which may be repeated at intervals—it becomes only necessary to inhale naturally with closed, and exhale with open, lips—an easy process, which I have ascertained practically may be kept up a long time, as, for example, for miles on a railroad-car, without much personal inconvenience. . . .

COMBINED CLOTHES BRUSH, FLASK, AND DRINKING CUP.

No. 490,964. Patented Jan. 31, 1893.

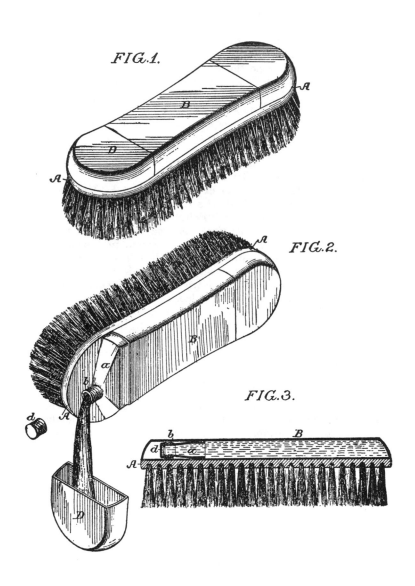

FIG.1.

FIG.2.

FIG.3.

Morning Pick-Me-Up Device

UNITED STATES PATENT OFFICE

COMBINED CLOTHES-BRUSH, FLASK, AND DRINKING-CUP

Specification forming part of Letters Patent No. 490,964, dated January 31, 1893
Application filed November 4, 1892. Serial No. 450,926. (No model)

. . . The object of my invention is to so construct a combined clothes brush, liquor flask, and drinking cup, as to provide for the compact disposal of the parts and thus attain the desired object without unduly increasing the size of the brush. . . .

Fitting snugly to the base of the contracted neck *a* of the flask is a cup D which, when applied to the neck of the flask as shown in Fig. 1, presents an outer surface flush with that of the flask and fitting snugly against the brush block, this cup conforming to the shape of that part of the block A which it covers, so that when the cup is in place on the flask the brush presents substantially the same appearance as an ordinary clothes brush, the cup, however, being readily removed when it is desired to use the same for drinking purposes, as shown in Fig. 2. . . .

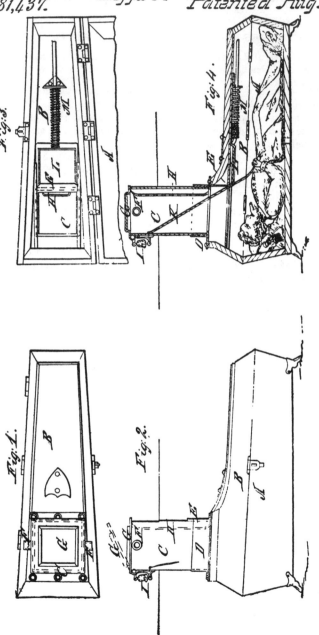

UNITED STATES PATENT OFFICE

IMPROVED BURIAL-CASE

Specification forming part of Letters Patent No. 81,437, dated August 25, 1868

. . . The nature of this invention consists in placing on the lid of the coffin, and directly over the face of the body laid therein, a square tube, which extends from the coffin up through and over the surface of the grave, said tube containing a ladder and a cord, one end of said cord being placed in the hand of the person laid in the coffin, and the other end of said cord being attached to a bell on the top of the square tube, so that, should a person be interred ere life is extinct, he can, on recovery to consciousness, ascend from the grave and the coffin by the ladder; or, if not able to ascend by said ladder, ring the bell, thereby giving an alarm, and thus save himself from premature burial and death; and if, on inspection, life is extinct, the tube is withdrawn, the sliding door closed, and the tube used for a similar purpose. . . .

1,303,851.

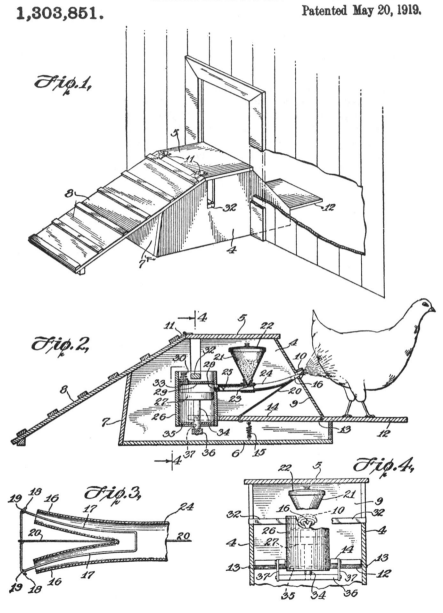

UNITED STATES PATENT OFFICE

POULTRY-DISINFECTOR

1,303,851

Specification of Letters Patent   Patented May 20, 1919

Application filed September 11, 1918. Serial No. 253,621

. . . This invention is a novel poultry disinfector, that is to say, an apparatus adapted to dust an insecticide powder, or to spray any other disinfectant upon live poultry. The principal object of the present invention is to afford such an apparatus which will be wholly self-acting and requiring very little attention from the poultry keeper, and which as well will be simple in structure, durable, effective in action and convenient of use. . . .

The illustrated embodiment of the present invention discloses a compact and portable apparatus which can be placed where desired, for example, in the entrance to the hen-house or a runway, and as will be seen it is adapted to operate in both directions, that is, on poultry coming and going so that they are sprayed at both front and rear. The apparatus preferably contains a box-like structure from the front of which projects a balanced low platform across or upon which a hen may walk or hop in entering or leaving the hen-house. Within the casing are the pneumatic sprayer directed outwardly at the front, and the mechanism for operating and controlling the same. . . .

By this arrangement, when the hen jumps on the front part 12 of the platform, the rear or interior end causes the piston to rise thus forcing air from the cylinder through the pipe 24, so as to carry the powder from the hopper 21 to the nozzle 16, whence it is sprayed upon the fowl. When the air is under pressure part of it will pass through the pinhole 23 into the hopper thus stirring up the disinfectant, and such air forced into the hopper promptly returns thus insuring passage of a suitable amount of powder into the pipe 24 at each operation. . . .

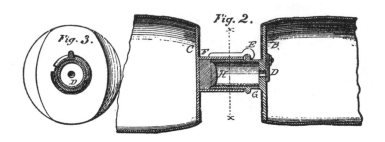

Fig. 2.

Fig. 3.

Fig. 1.

Life-Preserver Collar

UNITED STATES PATENT OFFICE

IMPROVEMENT IN LIFE-PRESERVERS

Letters Patent No. 100,906, dated March 15, 1870
The Schedule referred to in these Letters Patent and making part of the same.

. . . My invention relates to life-preservers, and consists of a circular hollow cylinder, made out of any light, flexible, air-tight material, of sufficient length to extend around the neck, and having attached to one of its ends a short tube with a hinged valve at its bottom, and to the other a similar tube of the proper size to slide within the former, and in providing these tubes with a bayonet-clasp. . . .

In using a life-preserver thus constructed, it is only necessary to remove the stopper H and fill the cylinder, by forcing air or gas through the tube G and past the valve D, and when filled to insert the stopper H, and fasten the cylinder about the neck.

The valve D should be so constructed and arranged that, with the aid of the pressure of air or gas against it within the cylinder, together with the stopper H at the end of the tube G, there can be no escape of the air or gas from the interior of the cylinder.

It is obvious that a device thus constructed, arranged, and applied, will keep the head of the wearer above the water, and thus saved from drowning. . . .

FISH LURE

Filed Nov. 19, 1927

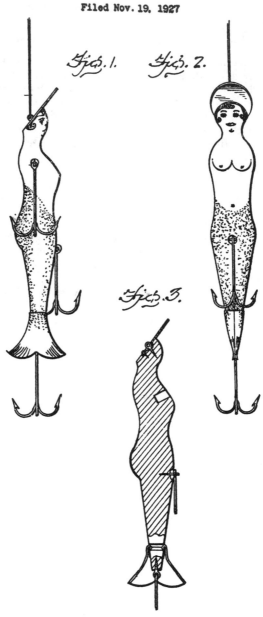

Fig. 1. Fig. 2.

Fig. 3.

Lorelei Bait

UNITED STATES PATENT OFFICE

DESIGN FOR A FISH LURE

Patented Mar. 20, 1928

Des. 74,759

Application filed November 19, 1927. Serial No. 24,209. Term of patent 7 years

... Be it known that I ... have invented a new, original, and ornamental Design for a Fish Lure, of which the following is a specification, reference being had to the accompanying drawing, forming part thereof.

Figure 1 is a side elevation of a fish lure, showing my new design.

Figure 2 is a front elevation thereof, and

Figure 3 is a longitudinal sectional view thereof. . . .

A CATALOG OF SELECTED
DOVER BOOKS
IN ALL FIELDS OF INTEREST

A CATALOG OF SELECTED DOVER BOOKS IN ALL FIELDS OF INTEREST

DRAWINGS OF REMBRANDT, edited by Seymour Slive. Updated Lippmann, Hofstede de Groot edition, with definitive scholarly apparatus. All portraits, biblical sketches, landscapes, nudes. Oriental figures, classical studies, together with selection of work by followers. 550 illustrations. Total of 630pp. 9⅛ × 12¼.
21485-0, 21486-9 Pa., Two-vol. set $29.90

GHOST AND HORROR STORIES OF AMBROSE BIERCE, Ambrose Bierce. 24 tales vividly imagined, strangely prophetic, and decades ahead of their time in technical skill: "The Damned Thing," "An Inhabitant of Carcosa," "The Eyes of the Panther," "Moxon's Master," and 20 more. 199pp. 5⅜ × 8½. 20767-6 Pa. $3.95

ETHICAL WRITINGS OF MAIMONIDES, Maimonides. Most significant ethical works of great medieval sage, newly translated for utmost precision, readability. Laws Concerning Character Traits, Eight Chapters, more. 192pp. 5⅜ × 8½.
24522-5 Pa. $4.50

THE EXPLORATION OF THE COLORADO RIVER AND ITS CANYONS, J. W. Powell. Full text of Powell's 1,000-mile expedition down the fabled Colorado in 1869. Superb account of terrain, geology, vegetation, Indians, famine, mutiny, treacherous rapids, mighty canyons, during exploration of last unknown part of continental U.S. 400pp. 5⅜ × 8½. 20094-9 Pa. $7.95

HISTORY OF PHILOSOPHY, Julián Marías. Clearest one-volume history on the market. Every major philosopher and dozens of others, to Existentialism and later. 505pp. 5⅜ × 8½. 21739-6 Pa. $9.95

ALL ABOUT LIGHTNING, Martin A. Uman. Highly readable non-technical survey of nature and causes of lightning, thunderstorms, ball lightning, St. Elmo's Fire, much more. Illustrated. 192pp. 5⅜ × 8½. 25237-X Pa. $5.95

SAILING ALONE AROUND THE WORLD, Captain Joshua Slocum. First man to sail around the world, alone, in small boat. One of great feats of seamanship told in delightful manner. 67 illustrations. 294pp. 5⅜ × 8½. 20326-3 Pa. $4.95

LETTERS AND NOTES ON THE MANNERS, CUSTOMS AND CONDITIONS OF THE NORTH AMERICAN INDIANS, George Catlin. Classic account of life among Plains Indians: ceremonies, hunt, warfare, etc. 312 plates. 572pp. of text. 6⅛ × 9¼. 22118-0, 22119-9, Pa. Two-vol. set $17.90

ALASKA: The Harriman Expedition, 1899, John Burroughs, John Muir, et al. Informative, engrossing accounts of two-month, 9,000-mile expedition. Native peoples, wildlife, forests, geography, salmon industry, glaciers, more. Profusely illustrated. 240 black-and-white line drawings. 124 black-and-white photographs. 3 maps. Index. 576pp. 5⅜ × 8½. 25109-8 Pa. $11.95

THE BOOK OF BEASTS: Being a Translation from a Latin Bestiary of the Twelfth Century, T. H. White. Wonderful catalog real and fanciful beasts: manticore, griffin, phoenix, amphivius, jaculus, many more. White's witty erudite commentary on scientific, historical aspects. Fascinating glimpse of medieval mind. Illustrated. 296pp. 5⅜ × 8¼. (Available in U.S. only) 24609-4 Pa. $6.95

FRANK LLOYD WRIGHT: ARCHITECTURE AND NATURE With 160 Illustrations, Donald Hoffmann. Profusely illustrated study of influence of nature—especially prairie—on Wright's designs for Fallingwater, Robie House, Guggenheim Museum, other masterpieces. 96pp. 9¼ × 10¾. 25098-9 Pa. $7.95

FRANK LLOYD WRIGHT'S FALLINGWATER, Donald Hoffmann. Wright's famous waterfall house: planning and construction of organic idea. History of site, owners, Wright's personal involvement. Photographs of various stages of building. Preface by Edgar Kaufmann, Jr. 100 illustrations. 112pp. 9¼ × 10.
23671-4 Pa. $8.95

YEARS WITH FRANK LLOYD WRIGHT: Apprentice to Genius, Edgar Tafel. Insightful memoir by a former apprentice presents a revealing portrait of Wright the man, the inspired teacher, the greatest American architect. 372 black-and-white illustrations. Preface. Index. vi + 228pp. 8¼ × 11. 24801-1 Pa. $10.95

THE STORY OF KING ARTHUR AND HIS KNIGHTS, Howard Pyle. Enchanting version of King Arthur fable has delighted generations with imaginative narratives of exciting adventures and unforgettable illustrations by the author. 41 illustrations. xviii + 313pp. 6⅛ × 9¼. 21445-1 Pa. $6.95

THE GODS OF THE EGYPTIANS, E. A. Wallis Budge. Thorough coverage of numerous gods of ancient Egypt by foremost Egyptologist. Information on evolution of cults, rites and gods; the cult of Osiris; the Book of the Dead and its rites; the sacred animals and birds; Heaven and Hell; and more. 956pp. 6⅛ × 9¼. 22055-9, 22056-7 Pa., Two-vol. set $21.90

A THEOLOGICO-POLITICAL TREATISE, Benedict Spinoza. Also contains unfinished *Political Treatise*. Great classic on religious liberty, theory of government on common consent. R. Elwes translation. Total of 421pp. 5⅜ × 8½.
20249-6 Pa. $6.95

INCIDENTS OF TRAVEL IN CENTRAL AMERICA, CHIAPAS, AND YUCATAN, John L. Stephens. Almost single-handed discovery of Maya culture; exploration of ruined cities, monuments, temples; customs of Indians. 115 drawings. 892pp. 5⅜ × 8½. 22404-X, 22405-8 Pa., Two-vol. set $15.90

LOS CAPRICHOS, Francisco Goya. 80 plates of wild, grotesque monsters and caricatures. Prado manuscript included. 183pp. 6⅜ × 9⅜. 22384-1 Pa. $5.95

AUTOBIOGRAPHY: The Story of My Experiments with Truth, Mohandas K. Gandhi. Not hagiography, but Gandhi in his own words. Boyhood, legal studies, purification, the growth of the Satyagraha (nonviolent protest) movement. Critical, inspiring work of the man who freed India. 480pp. 5⅜ × 8½. (Available in U.S. only) 24593-4 Pa. $6.95

ILLUSTRATED DICTIONARY OF HISTORIC ARCHITECTURE, edited by Cyril M. Harris. Extraordinary compendium of clear, concise definitions for over 5,000 important architectural terms complemented by over 2,000 line drawings. Covers full spectrum of architecture from ancient ruins to 20th-century Modernism. Preface. 592pp. 7½ × 9⅝. 24444-X Pa. $15.95

THE NIGHT BEFORE CHRISTMAS, Clement Moore. Full text, and woodcuts from original 1848 book. Also critical, historical material. 19 illustrations. 40pp. 4⅝ × 6. 22797-9 Pa. $2.50

THE LESSON OF JAPANESE ARCHITECTURE: 165 Photographs, Jiro Harada. Memorable gallery of 165 photographs taken in the 1930's of exquisite Japanese homes of the well-to-do and historic buildings. 13 line diagrams. 192pp. 8⅜ × 11¼. 24778-3 Pa. $10.95

THE AUTOBIOGRAPHY OF CHARLES DARWIN AND SELECTED LETTERS, edited by Francis Darwin. The fascinating life of eccentric genius composed of an intimate memoir by Darwin (intended for his children); commentary by his son, Francis; hundreds of fragments from notebooks, journals, papers; and letters to and from Lyell, Hooker, Huxley, Wallace and Henslow. xi + 365pp. 5⅜ × 8. 20479-0 Pa. $6.95

WONDERS OF THE SKY: Observing Rainbows, Comets, Eclipses, the Stars and Other Phenomena, Fred Schaaf. Charming, easy-to-read poetic guide to all manner of celestial events visible to the naked eye. Mock suns, glories, Belt of Venus, more. Illustrated. 299pp. 5¼ × 8¼. 24402-4 Pa. $7.95

BURNHAM'S CELESTIAL HANDBOOK, Robert Burnham, Jr. Thorough guide to the stars beyond our solar system. Exhaustive treatment. Alphabetical by constellation: Andromeda to Cetus in Vol. 1; Chamaeleon to Orion in Vol. 2; and Pavo to Vulpecula in Vol. 3. Hundreds of illustrations. Index in Vol. 3. 2,000pp. 6⅛ × 9¼. 23567-X, 23568-8, 23673-0 Pa., Three-vol. set $38.85

STAR NAMES: Their Lore and Meaning, Richard Hinckley Allen. Fascinating history of names various cultures have given to constellations and literary and folkloristic uses that have been made of stars. Indexes to subjects. Arabic and Greek names. Biblical references. Bibliography. 563pp. 5⅜ × 8½. 21079-0 Pa. $8.95

THIRTY YEARS THAT SHOOK PHYSICS: The Story of Quantum Theory, George Gamow. Lucid, accessible introduction to influential theory of energy and matter. Careful explanations of Dirac's anti-particles, Bohr's model of the atom, much more. 12 plates. Numerous drawings. 240pp. 5⅜ × 8½. 24895-X Pa. $5.95

CHINESE DOMESTIC FURNITURE IN PHOTOGRAPHS AND MEASURED DRAWINGS, Gustav Ecke. A rare volume, now affordably priced for antique collectors, furniture buffs and art historians. Detailed review of styles ranging from early Shang to late Ming. Unabridged republication. 161 black-and-white drawings, photos. Total of 224pp. 8⅜ × 11¼. (Available in U.S. only) 25171-3 Pa. $13.95

VINCENT VAN GOGH: A Biography, Julius Meier-Graefe. Dynamic, penetrating study of artist's life, relationship with brother, Theo, painting techniques, travels, more. Readable, engrossing. 160pp. 5⅜ × 8½. (Available in U.S. only) 25253-1 Pa. $4.95

HOW TO WRITE, Gertrude Stein. Gertrude Stein claimed anyone could understand her unconventional writing—here are clues to help. Fascinating improvisations, language experiments, explanations illuminate Stein's craft and the art of writing. Total of 414pp. 4⅝ × 6⅜. 23144-5 Pa. $6.95

ADVENTURES AT SEA IN THE GREAT AGE OF SAIL: Five Firsthand Narratives, edited by Elliot Snow. Rare true accounts of exploration, whaling, shipwreck, fierce natives, trade, shipboard life, more. 33 illustrations. Introduction. 353pp. 5⅜ × 8½. 25177-2 Pa. $8.95

THE HERBAL OR GENERAL HISTORY OF PLANTS, John Gerard. Classic descriptions of about 2,850 plants—with over 2,700 illustrations—includes Latin and English names, physical descriptions, varieties, time and place of growth, more. 2,706 illustrations. xlv + 1,678pp. 8½ × 12¼. 23147-X Cloth. $75.00

DOROTHY AND THE WIZARD IN OZ, L. Frank Baum. Dorothy and the Wizard visit the center of the Earth, where people are vegetables, glass houses grow and Oz characters reappear. Classic sequel to *Wizard of Oz.* 256pp. 5⅜ × 8. 24714-7 Pa. $4.95

SONGS OF EXPERIENCE: Facsimile Reproduction with 26 Plates in Full Color, William Blake. This facsimile of Blake's original "Illuminated Book" reproduces 26 full-color plates from a rare 1826 edition. Includes "The Tyger," "London," "Holy Thursday," and other immortal poems. 26 color plates. Printed text of poems. 48pp. 5¼ × 7. 24636-1 Pa. $3.50

SONGS OF INNOCENCE, William Blake. The first and most popular of Blake's famous "Illuminated Books," in a facsimile edition reproducing all 31 brightly colored plates. Additional printed text of each poem. 64pp. 5¼ × 7. 22764-2 Pa. $3.50

PRECIOUS STONES, Max Bauer. Classic, thorough study of diamonds, rubies, emeralds, garnets, etc.: physical character, occurrence, properties, use, similar topics. 20 plates, 8 in color. 94 figures. 659pp. 6⅛ × 9¼. 21910-0, 21911-9 Pa., Two-vol. set $15.90

ENCYCLOPEDIA OF VICTORIAN NEEDLEWORK, S. F. A. Caulfeild and Blanche Saward. Full, precise descriptions of stitches, techniques for dozens of needlecrafts—most exhaustive reference of its kind. Over 800 figures. Total of 679pp. 8⅛ × 11. Two volumes. Vol. 1 22800-2 Pa. $11.95
Vol. 2 22801-0 Pa. $11.95

THE MARVELOUS LAND OF OZ, L. Frank Baum. Second Oz book, the Scarecrow and Tin Woodman are back with hero named Tip, Oz magic. 136 illustrations. 287pp. 5⅜ × 8½. 20692-0 Pa. $5.95

WILD FOWL DECOYS, Joel Barber. Basic book on the subject, by foremost authority and collector. Reveals history of decoy making and rigging, place in American culture, different kinds of decoys, how to make them, and how to use them. 140 plates. 156pp. 7⅞ × 10¾. 20011-6 Pa. $8.95

HISTORY OF LACE, Mrs. Bury Palliser. Definitive, profusely illustrated chronicle of lace from earliest times to late 19th century. Laces of Italy, Greece, England, France, Belgium, etc. Landmark of needlework scholarship. 266 illustrations. 672pp. 6⅛ × 9¼. 24742-2 Pa. $14.95

ILLUSTRATED GUIDE TO SHAKER FURNITURE, Robert Meader. All furniture and appurtenances, with much on unknown local styles. 235 photos. 146pp. 9 × 12. 22819-3 Pa. $8.95

WHALE SHIPS AND WHALING: A Pictorial Survey, George Francis Dow. Over 200 vintage engravings, drawings, photographs of barks, brigs, cutters, other vessels. Also harpoons, lances, whaling guns, many other artifacts. Comprehensive text by foremost authority. 207 black-and-white illustrations. 288pp. 6 × 9. 24808-9 Pa. $8.95

THE BERTRAMS, Anthony Trollope. Powerful portrayal of blind self-will and thwarted ambition includes one of Trollope's most heartrending love stories. 497pp. 5⅜ × 8½. 25119-5 Pa. $9.95

ADVENTURES WITH A HAND LENS, Richard Headstrom. Clearly written guide to observing and studying flowers and grasses, fish scales, moth and insect wings, egg cases, buds, feathers, seeds, leaf scars, moss, molds, ferns, common crystals, etc.—all with an ordinary, inexpensive magnifying glass. 209 exact line drawings aid in your discoveries. 220pp. 5⅜ × 8½. 23330-8 Pa. $4.95

RODIN ON ART AND ARTISTS, Auguste Rodin. Great sculptor's candid, wide-ranging comments on meaning of art; great artists; relation of sculpture to poetry, painting, music; philosophy of life, more. 76 superb black-and-white illustrations of Rodin's sculpture, drawings and prints. 119pp. 8⅜ × 11¼. 24487-3 Pa. $7.95

FIFTY CLASSIC FRENCH FILMS, 1912–1982: A Pictorial Record, Anthony Slide. Memorable stills from Grand Illusion, Beauty and the Beast, Hiroshima, Mon Amour, many more. Credits, plot synopses, reviews, etc. 160pp. 8¼ × 11. 25256-6 Pa. $11.95

THE PRINCIPLES OF PSYCHOLOGY, William James. Famous long course complete, unabridged. Stream of thought, time perception, memory, experimental methods; great work decades ahead of its time. 94 figures. 1,391pp. 5⅜ × 8½. 20381-6, 20382-4 Pa., Two-vol. set $23.90

BODIES IN A BOOKSHOP, R. T. Campbell. Challenging mystery of blackmail and murder with ingenious plot and superbly drawn characters. In the best tradition of British suspense fiction. 192pp. 5⅜ × 8½. 24720-1 Pa. $3.95

CALLAS: PORTRAIT OF A PRIMA DONNA, George Jellinek. Renowned commentator on the musical scene chronicles incredible career and life of the most controversial, fascinating, influential operatic personality of our time. 64 black-and-white photographs. 416pp. 5⅜ × 8¼. 25047-4 Pa. $8.95

GEOMETRY, RELATIVITY AND THE FOURTH DIMENSION, Rudolph Rucker. Exposition of fourth dimension, concepts of relativity as Flatland characters continue adventures. Popular, easily followed yet accurate, profound. 141 illustrations. 133pp. 5⅜ × 8½. 23400-2 Pa. $3.95

HOUSEHOLD STORIES BY THE BROTHERS GRIMM, with pictures by Walter Crane. 53 classic stories—Rumpelstiltskin, Rapunzel, Hansel and Gretel, the Fisherman and his Wife, Snow White, Tom Thumb, Sleeping Beauty, Cinderella, and so much more—lavishly illustrated with original 19th century drawings. 114 illustrations. x + 269pp. 5⅜ × 8½. 21080-4 Pa. $4.95

SUNDIALS, Albert Waugh. Far and away the best, most thorough coverage of ideas, mathematics concerned, types, construction, adjusting anywhere. Over 100 illustrations. 230pp. 5⅜ × 8½. 22947-5 Pa. $4.95

PICTURE HISTORY OF THE NORMANDIE: With 190 Illustrations, Frank O. Braynard. Full story of legendary French ocean liner: Art Deco interiors, design innovations, furnishings, celebrities, maiden voyage, tragic fire, much more. Extensive text. 144pp. 8⅜ × 11¾. 25257-4 Pa. $10.95

THE FIRST AMERICAN COOKBOOK: A Facsimile of "American Cookery," 1796, Amelia Simmons. Facsimile of the first American-written cookbook published in the United States contains authentic recipes for colonial favorites—pumpkin pudding, winter squash pudding, spruce beer, Indian slapjacks, and more. Introductory Essay and Glossary of colonial cooking terms. 80pp. 5⅜ × 8½. 24710-4 Pa. $3.50

101 PUZZLES IN THOUGHT AND LOGIC, C. R. Wylie, Jr. Solve murders and robberies, find out which fishermen are liars, how a blind man could possibly identify a color—purely by your own reasoning! 107pp. 5⅜ × 8½. 20367-0 Pa. $2.50

THE BOOK OF WORLD-FAMOUS MUSIC—CLASSICAL, POPULAR AND FOLK, James J. Fuld. Revised and enlarged republication of landmark work in musico-bibliography. Full information about nearly 1,000 songs and compositions including first lines of music and lyrics. New supplement. Index. 800pp. 5⅜ × 8¼. 24857-7 Pa. $15.95

ANTHROPOLOGY AND MODERN LIFE, Franz Boas. Great anthropologist's classic treatise on race and culture. Introduction by Ruth Bunzel. Only inexpensive paperback edition. 255pp. 5⅜ × 8½. 25245-0 Pa. $6.95

THE TALE OF PETER RABBIT, Beatrix Potter. The inimitable Peter's terrifying adventure in Mr. McGregor's garden, with all 27 wonderful, full-color Potter illustrations. 55pp. 4¼ × 5½. (Available in U.S. only) 22827-4 Pa. $1.75

THREE PROPHETIC SCIENCE FICTION NOVELS, H. G. Wells. *When the Sleeper Wakes, A Story of the Days to Come* and *The Time Machine* (full version). 335pp. 5⅜ × 8½. (Available in U.S. only) 20605-X Pa. $6.95

APICIUS COOKERY AND DINING IN IMPERIAL ROME, edited and translated by Joseph Dommers Vehling. Oldest known cookbook in existence offers readers a clear picture of what foods Romans ate, how they prepared them, etc. 49 illustrations. 301pp. 6⅛ × 9¼. 23563-7 Pa. $7.95

SHAKESPEARE LEXICON AND QUOTATION DICTIONARY, Alexander Schmidt. Full definitions, locations, shades of meaning of every word in plays and poems. More than 50,000 exact quotations. 1,485pp. 6½ × 9¼. 22726-X, 22727-8 Pa., Two-vol. set $29.90

THE WORLD'S GREAT SPEECHES, edited by Lewis Copeland and Lawrence W. Lamm. Vast collection of 278 speeches from Greeks to 1970. Powerful and effective models; unique look at history. 842pp. 5⅜ × 8½. 20468-5 Pa. $11.95

CATALOG OF DOVER BOOKS

THE BLUE FAIRY BOOK, Andrew Lang. The first, most famous collection, with many familiar tales: Little Red Riding Hood, Aladdin and the Wonderful Lamp, Puss in Boots, Sleeping Beauty, Hansel and Gretel, Rumpelstiltskin; 37 in all. 138 illustrations. 390pp. 5⅜ × 8½. 21437-0 Pa. $6.95

THE STORY OF THE CHAMPIONS OF THE ROUND TABLE, Howard Pyle. Sir Launcelot, Sir Tristram and Sir Percival in spirited adventures of love and triumph retold in Pyle's inimitable style. 50 drawings, 31 full-page. xviii + 329pp. 6½ × 9¼. 21883-X Pa. $7.95

AUDUBON AND HIS JOURNALS, Maria Audubon. Unmatched two-volume portrait of the great artist, naturalist and author contains his journals, an excellent biography by his granddaughter, expert annotations by the noted ornithologist, Dr. Elliott Coues, and 37 superb illustrations. Total of 1,200pp. 5⅜ × 8.
Vol. I 25143-8 Pa. $8.95
Vol. II 25144-6 Pa. $8.95

GREAT DINOSAUR HUNTERS AND THEIR DISCOVERIES, Edwin H. Colbert. Fascinating, lavishly illustrated chronicle of dinosaur research, 1820's to 1960. Achievements of Cope, Marsh, Brown, Buckland, Mantell, Huxley, many others. 384pp. 5¼ × 8¼. 24701-5 Pa. $7.95

THE TASTEMAKERS, Russell Lynes. Informal, illustrated social history of American taste 1850's–1950's. First popularized categories Highbrow, Lowbrow, Middlebrow. 129 illustrations. New (1979) afterword. 384pp. 6 × 9.
23993-4 Pa. $8.95

DOUBLE CROSS PURPOSES, Ronald A. Knox. A treasure hunt in the Scottish Highlands, an old map, unidentified corpse, surprise discoveries keep reader guessing in this cleverly intricate tale of financial skullduggery. 2 black-and-white maps. 320pp. 5⅜ × 8½. (Available in U.S. only) 25032-6 Pa. $6.95

AUTHENTIC VICTORIAN DECORATION AND ORNAMENTATION IN FULL COLOR: 46 Plates from "Studies in Design," Christopher Dresser. Superb full-color lithographs reproduced from rare original portfolio of a major Victorian designer. 48pp. 9¼ × 12¼. 25083-0 Pa. $7.95

PRIMITIVE ART, Franz Boas. Remains the best text ever prepared on subject, thoroughly discussing Indian, African, Asian, Australian, and, especially, Northern American primitive art. Over 950 illustrations show ceramics, masks, totem poles, weapons, textiles, paintings, much more. 376pp. 5⅜ × 8. 20025-6 Pa. $6.95

SIDELIGHTS ON RELATIVITY, Albert Einstein. Unabridged republication of two lectures delivered by the great physicist in 1920–21. *Ether and Relativity* and *Geometry and Experience*. Elegant ideas in non-mathematical form, accessible to intelligent layman. vi + 56pp. 5⅜ × 8½. 24511-X Pa. $2.95

THE WIT AND HUMOR OF OSCAR WILDE, edited by Alvin Redman. More than 1,000 ripostes, paradoxes, wisecracks: Work is the curse of the drinking classes, I can resist everything except temptation, etc. 258pp. 5⅜ × 8½. 20602-5 Pa. $4.95

ADVENTURES WITH A MICROSCOPE, Richard Headstrom. 59 adventures with clothing fibers, protozoa, ferns and lichens, roots and leaves, much more. 142 illustrations. 232pp. 5⅜ × 8½. 23471-1 Pa. $3.95

PLANTS OF THE BIBLE, Harold N. Moldenke and Alma L. Moldenke. Standard reference to all 230 plants mentioned in Scriptures. Latin name, biblical reference, uses, modern identity, much more. Unsurpassed encyclopedic resource for scholars, botanists, nature lovers, students of Bible. Bibliography. Indexes. 123 black-and-white illustrations. 384pp. 6 × 9. 25069-5 Pa. $8.95

FAMOUS AMERICAN WOMEN: A Biographical Dictionary from Colonial Times to the Present, Robert McHenry, ed. From Pocahontas to Rosa Parks, 1,035 distinguished American women documented in separate biographical entries. Accurate, up-to-date data, numerous categories, spans 400 years. Indices. 493pp. 6½ × 9¼. 24523-3 Pa. $10.95

THE FABULOUS INTERIORS OF THE GREAT OCEAN LINERS IN HISTORIC PHOTOGRAPHS, William H. Miller, Jr. Some 200 superb photographs capture exquisite interiors of world's great "floating palaces"—1890's to 1980's: *Titanic, Ile de France, Queen Elizabeth, United States, Europa,* more. Approx. 200 black-and-white photographs. Captions. Text. Introduction. 160pp. 8⅞ × 11¾. 24756-2 Pa. $9.95

THE GREAT LUXURY LINERS, 1927–1954: A Photographic Record, William H. Miller, Jr. Nostalgic tribute to heyday of ocean liners. 186 photos of Ile de France, Normandie, Leviathan, Queen Elizabeth, United States, many others. Interior and exterior views. Introduction. Captions. 160pp. 9 × 12. 24056-8 Pa. $10.95

A NATURAL HISTORY OF THE DUCKS, John Charles Phillips. Great landmark of ornithology offers complete detailed coverage of nearly 200 species and subspecies of ducks: gadwall, sheldrake, merganser, pintail, many more. 74 full-color plates, 102 black-and-white. Bibliography. Total of 1,920pp. 8⅝ × 11¼. 25141-1, 25142-X Cloth. Two-vol. set $100.00

THE SEAWEED HANDBOOK: An Illustrated Guide to Seaweeds from North Carolina to Canada, Thomas F. Lee. Concise reference covers 78 species. Scientific and common names, habitat, distribution, more. Finding keys for easy identification. 224pp. 5⅜ × 8½. 25215-9 Pa. $6.95

THE TEN BOOKS OF ARCHITECTURE: The 1755 Leoni Edition, Leon Battista Alberti. Rare classic helped introduce the glories of ancient architecture to the Renaissance. 68 black-and-white plates. 336pp. 8⅜ × 11¼. 25239-6 Pa. $14.95

MISS MACKENZIE, Anthony Trollope. Minor masterpieces by Victorian master unmasks many truths about life in 19th-century England. First inexpensive edition in years. 392pp. 5⅜ × 8½. 25201-9 Pa. $8.95

THE RIME OF THE ANCIENT MARINER, Gustave Doré, Samuel Taylor Coleridge. Dramatic engravings considered by many to be his greatest work. The terrifying space of the open sea, the storms and whirlpools of an unknown ocean, the ice of Antarctica, more—all rendered in a powerful, chilling manner. Full text. 38 plates. 77pp. 9¼ × 12. 22305-1 Pa. $4.95

THE EXPEDITIONS OF ZEBULON MONTGOMERY PIKE, Zebulon Montgomery Pike. Fascinating first-hand accounts (1805–6) of exploration of Mississippi River, Indian wars, capture by Spanish dragoons, much more. 1,088pp. 5⅜ × 8½. 25254-X, 25255-8 Pa. Two-vol. set $25.90

A CONCISE HISTORY OF PHOTOGRAPHY: Third Revised Edition, Helmut Gernsheim. Best one-volume history—camera obscura, photochemistry, daguerreotypes, evolution of cameras, film, more. Also artistic aspects—landscape, portraits, fine art, etc. 281 black-and-white photographs. 26 in color. 176pp. 8⅜ × 11¼. 25128-4 Pa. $13.95

THE DORÉ BIBLE ILLUSTRATIONS, Gustave Doré. 241 detailed plates from the Bible: the Creation scenes, Adam and Eve, Flood, Babylon, battle sequences, life of Jesus, etc. Each plate is accompanied by the verses from the King James version of the Bible. 241pp. 9 × 12. 23004-X Pa. $9.95

HUGGER-MUGGER IN THE LOUVRE, Elliot Paul. Second Homer Evans mystery-comedy. Theft at the Louvre involves sleuth in hilarious, madcap caper. "A knockout."—Books. 336pp. 5⅜ × 8½. 25185-3 Pa. $5.95

FLATLAND, E. A. Abbott. Intriguing and enormously popular science-fiction classic explores the complexities of trying to survive as a two-dimensional being in a three-dimensional world. Amusingly illustrated by the author. 16 illustrations. 103pp. 5⅜ × 8½. 20001-9 Pa. $2.50

THE HISTORY OF THE LEWIS AND CLARK EXPEDITION, Meriwether Lewis and William Clark, edited by Elliott Coues. Classic edition of Lewis and Clark's day-by-day journals that later became the basis for U.S. claims to Oregon and the West. Accurate and invaluable geographical, botanical, biological, meteorological and anthropological material. Total of 1,508pp. 5⅜ × 8½.
21268-8, 21269-6, 21270-X Pa. Three-vol. set $26.85

LANGUAGE, TRUTH AND LOGIC, Alfred J. Ayer. Famous, clear introduction to Vienna, Cambridge schools of Logical Positivism. Role of philosophy, elimination of metaphysics, nature of analysis, etc. 160pp. 5⅜ × 8½. (Available in U.S. and Canada only) 20010-8 Pa. $3.95

MATHEMATICS FOR THE NONMATHEMATICIAN, Morris Kline. Detailed, college-level treatment of mathematics in cultural and historical context, with numerous exercises. For liberal arts students. Preface. Recommended Reading Lists. Tables. Index. Numerous black-and-white figures. xvi + 641pp. 5⅜ × 8½.
24823-2 Pa. $11.95

HANDBOOK OF PICTORIAL SYMBOLS, Rudolph Modley. 3,250 signs and symbols, many systems in full; official or heavy commercial use. Arranged by subject. Most in Pictorial Archive series. 143pp. 8⅛ × 11. 23357-X Pa. $6.95

INCIDENTS OF TRAVEL IN YUCATAN, John L. Stephens. Classic (1843) exploration of jungles of Yucatan, looking for evidences of Maya civilization. Travel adventures, Mexican and Indian culture, etc. Total of 669pp. 5⅜ × 8½.
20926-1, 20927-X Pa., Two-vol. set $11.90

DEGAS: An Intimate Portrait, Ambroise Vollard. Charming, anecdotal memoir by famous art dealer of one of the greatest 19th-century French painters. 14 black-and-white illustrations. Introduction by Harold L. Van Doren. 96pp. 5⅜ × 8½.
25131-4 Pa. $4.95

PERSONAL NARRATIVE OF A PILGRIMAGE TO ALMANDINAH AND MECCAH, Richard Burton. Great travel classic by remarkably colorful personality. Burton, disguised as a Moroccan, visited sacred shrines of Islam, narrowly escaping death. 47 illustrations. 959pp. 5⅜ × 8½.   21217-3, 21218-1 Pa., Two-vol. set $19.90

PHRASE AND WORD ORIGINS, A. H. Holt. Entertaining, reliable, modern study of more than 1,200 colorful words, phrases, origins and histories. Much unexpected information. 254pp. 5⅜ × 8½.   20758-7 Pa. $5.95

THE RED THUMB MARK, R. Austin Freeman. In this first Dr. Thorndyke case, the great scientific detective draws fascinating conclusions from the nature of a single fingerprint. Exciting story, authentic science. 320pp. 5⅜ × 8½. (Available in U.S. only)   25210-8 Pa. $6.95

AN EGYPTIAN HIEROGLYPHIC DICTIONARY, E. A. Wallis Budge. Monumental work containing about 25,000 words or terms that occur in texts ranging from 3000 B.C. to 600 A.D. Each entry consists of a transliteration of the word, the word in hieroglyphs, and the meaning in English. 1,314pp. 6⅜ × 10.
23615-3, 23616-1 Pa., Two-vol. set $31.90

THE COMPLEAT STRATEGYST: Being a Primer on the Theory of Games of Strategy, J. D. Williams. Highly entertaining classic describes, with many illustrated examples, how to select best strategies in conflict situations. Prefaces. Appendices. xvi + 268pp. 5⅜ × 8½.   25101-2 Pa. $5.95

THE ROAD TO OZ, L. Frank Baum. Dorothy meets the Shaggy Man, little Button-Bright and the Rainbow's beautiful daughter in this delightful trip to the magical Land of Oz. 272pp. 5⅜ × 8.   25208-6 Pa. $5.95

POINT AND LINE TO PLANE, Wassily Kandinsky. Seminal exposition of role of point, line, other elements in non-objective painting. Essential to understanding 20th-century art. 127 illustrations. 192pp. 6½ × 9¼.   23808-3 Pa. $4.95

LADY ANNA, Anthony Trollope. Moving chronicle of Countess Lovel's bitter struggle to win for herself and daughter Anna their rightful rank and fortune—perhaps at cost of sanity itself. 384pp. 5⅜ × 8½.   24669-8 Pa. $8.95

EGYPTIAN MAGIC, E. A. Wallis Budge. Sums up all that is known about magic in Ancient Egypt: the role of magic in controlling the gods, powerful amulets that warded off evil spirits, scarabs of immortality, use of wax images, formulas and spells, the secret name, much more. 253pp. 5⅜ × 8½.   22681-6 Pa. $4.50

THE DANCE OF SIVA, Ananda Coomaraswamy. Preeminent authority unfolds the vast metaphysic of India: the revelation of her art, conception of the universe, social organization, etc. 27 reproductions of art masterpieces. 192pp. 5⅜ × 8½.
24817-8 Pa. $5.95

CHRISTMAS CUSTOMS AND TRADITIONS, Clement A. Miles. Origin, evolution, significance of religious, secular practices. Caroling, gifts, yule logs, much more. Full, scholarly yet fascinating; non-sectarian. 400pp. 5⅜ × 8½.
23354-5 Pa. $6.95

THE HUMAN FIGURE IN MOTION, Eadweard Muybridge. More than 4,500 stopped-action photos, in action series, showing undraped men, women, children jumping, lying down, throwing, sitting, wrestling, carrying, etc. 390pp. 7⅞ × 10⅝.
20204-6 Cloth. $21.95

THE MAN WHO WAS THURSDAY, Gilbert Keith Chesterton. Witty, fast-paced novel about a club of anarchists in turn-of-the-century London. Brilliant social, religious, philosophical speculations. 128pp. 5⅜ × 8½.
25121-7 Pa. $3.95

A CEZANNE SKETCHBOOK: Figures, Portraits, Landscapes and Still Lifes, Paul Cezanne. Great artist experiments with tonal effects, light, mass, other qualities in over 100 drawings. A revealing view of developing master painter, precursor of Cubism. 102 black-and-white illustrations. 144pp. 8¾ × 6⅜.
24790-2 Pa. $5.95

AN ENCYCLOPEDIA OF BATTLES: Accounts of Over 1,560 Battles from 1479 B.C. to the Present, David Eggenberger. Presents essential details of every major battle in recorded history, from the first battle of Megiddo in 1479 B.C. to Grenada in 1984. List of Battle Maps. New Appendix covering the years 1967–1984. Index. 99 illustrations. 544pp. 6½ × 9¼.
24913-1 Pa. $14.95

AN ETYMOLOGICAL DICTIONARY OF MODERN ENGLISH, Ernest Weekley. Richest, fullest work, by foremost British lexicographer. Detailed word histories. Inexhaustible. Total of 856pp. 6½ × 9¼.
21873-2, 21874-0 Pa., Two-vol. set $17.00

WEBSTER'S AMERICAN MILITARY BIOGRAPHIES, edited by Robert McHenry. Over 1,000 figures who shaped 3 centuries of American military history. Detailed biographies of Nathan Hale, Douglas MacArthur, Mary Hallaren, others. Chronologies of engagements, more. Introduction. Addenda. 1,033 entries in alphabetical order. xi + 548pp. 6½ × 9¼. (Available in U.S. only)
24758-9 Pa. $13.95

LIFE IN ANCIENT EGYPT, Adolf Erman. Detailed older account, with much not in more recent books: domestic life, religion, magic, medicine, commerce, and whatever else needed for complete picture. Many illustrations. 597pp. 5⅜ × 8½.
22632-8 Pa. $8.95

HISTORIC COSTUME IN PICTURES, Braun & Schneider. Over 1,450 costumed figures shown, covering a wide variety of peoples: kings, emperors, nobles, priests, servants, soldiers, scholars, townsfolk, peasants, merchants, courtiers, cavaliers, and more. 256pp. 8⅜ × 11¼.
23150-X Pa. $9.95

THE NOTEBOOKS OF LEONARDO DA VINCI, edited by J. P. Richter. Extracts from manuscripts reveal great genius; on painting, sculpture, anatomy, sciences, geography, etc. Both Italian and English. 186 ms. pages reproduced, plus 500 additional drawings, including studies for *Last Supper, Sforza* monument, etc. 860pp. 7⅞ × 10¾. (Available in U.S. only) 22572-0, 22573-9 Pa., Two-vol. set $31.90

THE ART NOUVEAU STYLE BOOK OF ALPHONSE MUCHA: All 72 Plates from "Documents Decoratifs" in Original Color, Alphonse Mucha. Rare copyright-free design portfolio by high priest of Art Nouveau. Jewelry, wallpaper, stained glass, furniture, figure studies, plant and animal motifs, etc. Only complete one-volume edition. 80pp. 9⅜ × 12¼. 24044-4 Pa. $9.95

ANIMALS: 1,419 COPYRIGHT-FREE ILLUSTRATIONS OF MAMMALS, BIRDS, FISH, INSECTS, ETC., edited by Jim Harter. Clear wood engravings present, in extremely lifelike poses, over 1,000 species of animals. One of the most extensive pictorial sourcebooks of its kind. Captions. Index. 284pp. 9 × 12.
23766-4 Pa. $9.95

OBELISTS FLY HIGH, C. Daly King. Masterpiece of American detective fiction, long out of print, involves murder on a 1935 transcontinental flight—"a very thrilling story"—NY Times. Unabridged and unaltered republication of the edition published by William Collins Sons & Co. Ltd., London, 1935. 288pp. 5⅜ × 8½. (Available in U.S. only) 25036-9 Pa. $5.95

VICTORIAN AND EDWARDIAN FASHION: A Photographic Survey, Alison Gernsheim. First fashion history completely illustrated by contemporary photographs. Full text plus 235 photos, 1840–1914, in which many celebrities appear. 240pp. 6½ × 9¼. 24205-6 Pa. $6.95

THE ART OF THE FRENCH ILLUSTRATED BOOK, 1700–1914, Gordon N. Ray. Over 630 superb book illustrations by Fragonard, Delacroix, Daumier, Doré, Grandville, Manet, Mucha, Steinlen, Toulouse-Lautrec and many others. Preface. Introduction. 633 halftones. Indices of artists, authors & titles, binders and provenances. Appendices. Bibliography. 608pp. 8⅜ × 11¼. 25086-5 Pa. $24.95

THE WONDERFUL WIZARD OF OZ, L. Frank Baum. Facsimile in full color of America's finest children's classic. 143 illustrations by W. W. Denslow. 267pp. 5⅜ × 8½. 20691-2 Pa. $7.95

FRONTIERS OF MODERN PHYSICS: New Perspectives on Cosmology, Relativity, Black Holes and Extraterrestrial Intelligence, Tony Rothman, et al. For the intelligent layman. Subjects include: cosmological models of the universe; black holes; the neutrino; the search for extraterrestrial intelligence. Introduction. 46 black-and-white illustrations. 192pp. 5⅜ × 8½. 24587-X Pa. $7.95

THE FRIENDLY STARS, Martha Evans Martin & Donald Howard Menzel. Classic text marshalls the stars together in an engaging, non-technical survey, presenting them as sources of beauty in night sky. 23 illustrations. Foreword. 2 star charts. Index. 147pp. 5⅜ × 8½. 21099-5 Pa. $3.95

FADS AND FALLACIES IN THE NAME OF SCIENCE, Martin Gardner. Fair, witty appraisal of cranks, quacks, and quackeries of science and pseudoscience: hollow earth, Velikovsky, orgone energy, Dianetics, flying saucers, Bridey Murphy, food and medical fads, etc. Revised, expanded In the Name of Science. "A very able and even-tempered presentation."—The New Yorker. 363pp. 5⅜ × 8.

20394-8 Pa. $6.95

ANCIENT EGYPT: ITS CULTURE AND HISTORY, J. E Manchip White. From pre-dynastics through Ptolemies: society, history, political structure, religion, daily life, literature, cultural heritage. 48 plates. 217pp. 5⅜ × 8½. 22548-8 Pa. $5.95

CATALOG OF DOVER BOOKS

SIR HARRY HOTSPUR OF HUMBLETHWAITE, Anthony Trollope. Incisive, unconventional psychological study of a conflict between a wealthy baronet, his idealistic daughter, and their scapegrace cousin. The 1870 novel in its first inexpensive edition in years. 250pp. 5⅜ × 8½. 24953-0 Pa. $5.95

LASERS AND HOLOGRAPHY, Winston E. Kock. Sound introduction to burgeoning field, expanded (1981) for second edition. Wave patterns, coherence, lasers, diffraction, zone plates, properties of holograms, recent advances. 84 illustrations. 160pp. 5⅜ × 8¼. (Except in United Kingdom) 24041-X Pa. $3.95

INTRODUCTION TO ARTIFICIAL INTELLIGENCE: SECOND, EN-LARGED EDITION, Philip C. Jackson, Jr. Comprehensive survey of artificial intelligence—the study of how machines (computers) can be made to act intelligently. Includes introductory and advanced material. Extensive notes updating the main text. 132 black-and-white illustrations. 512pp. 5⅜ × 8½. 24864-X Pa. $8.95

HISTORY OF INDIAN AND INDONESIAN ART, Ananda K. Coomaraswamy. Over 400 illustrations illuminate classic study of Indian art from earliest Harappa finds to early 20th century. Provides philosophical, religious and social insights. 304pp. 6⅜ × 9¾. 25005-9 Pa. $9.95

THE GOLEM, Gustav Meyrink. Most famous supernatural novel in modern European literature, set in Ghetto of Old Prague around 1890. Compelling story of mystical experiences, strange transformations, profound terror. 13 black-and-white illustrations. 224pp. 5⅜ × 8½. (Available in U.S. only) 25025-3 Pa. $6.95

ARMADALE, Wilkie Collins. Third great mystery novel by the author of *The Woman in White* and *The Moonstone*. Original magazine version with 40 illustrations. 597pp. 5⅜ × 8½. 23429-0 Pa. $9.95

PICTORIAL ENCYCLOPEDIA OF HISTORIC ARCHITECTURAL PLANS, DETAILS AND ELEMENTS: With 1,880 Line Drawings of Arches, Domes, Doorways, Facades, Gables, Windows, etc., John Theodore Haneman. Sourcebook of inspiration for architects, designers, others. Bibliography. Captions. 141pp. 9 × 12. 24605-1 Pa. $7.95

BENCHLEY LOST AND FOUND, Robert Benchley. Finest humor from early 30's, about pet peeves, child psychologists, post office and others. Mostly unavailable elsewhere. 73 illustrations by Peter Arno and others. 183pp. 5⅜ × 8½. 22410-4 Pa. $4.95

ERTÉ GRAPHICS, Erté. Collection of striking color graphics: *Seasons, Alphabet, Numerals, Aces* and *Precious Stones*. 50 plates, including 4 on covers. 48pp. 9⅜ × 12¼. 23580-7 Pa. $6.95

THE JOURNAL OF HENRY D. THOREAU, edited by Bradford Torrey, F. H. Allen. Complete reprinting of 14 volumes, 1837–61, over two million words; the sourcebooks for *Walden*, etc. Definitive. All original sketches, plus 75 photographs. 1,804pp. 8½ × 12¼. 20312-3, 20313-1 Cloth., Two-vol. set $120.00

CASTLES: THEIR CONSTRUCTION AND HISTORY, Sidney Toy. Traces castle development from ancient roots. Nearly 200 photographs and drawings illustrate moats, keeps, baileys, many other features. Caernarvon, Dover Castles, Hadrian's Wall, Tower of London, dozens more. 256pp. 5⅜ × 8¼. 24898-4 Pa. $6.95

AMERICAN CLIPPER SHIPS: 1833–1858, Octavius T. Howe & Frederick C. Matthews. Fully-illustrated, encyclopedic review of 352 clipper ships from the period of America's greatest maritime supremacy. Introduction. 109 halftones. 5 black-and-white line illustrations. Index. Total of 928pp. 5⅜ × 8½.
25115-2, 25116-0 Pa., Two-vol. set $17.90

TOWARDS A NEW ARCHITECTURE, Le Corbusier. Pioneering manifesto by great architect, near legendary founder of "International School." Technical and aesthetic theories, views on industry, economics, relation of form to function, "mass-production spirit," much more. Profusely illustrated. Unabridged translation of 13th French edition. Introduction by Frederick Etchells. 320pp. 6⅛ × 9¼. (Available in U.S. only) 25023-7 Pa. $8.95

THE BOOK OF KELLS, edited by Blanche Cirker. Inexpensive collection of 32 full-color, full-page plates from the greatest illuminated manuscript of the Middle Ages, painstakingly reproduced from rare facsimile edition. Publisher's Note. Captions. 32pp. 9⅜ × 12¼. 24345-1 Pa. $4.95

BEST SCIENCE FICTION STORIES OF H. G. WELLS, H. G. Wells. Full novel *The Invisible Man,* plus 17 short stories: "The Crystal Egg," "Aepyornis Island," "The Strange Orchid," etc. 303pp. 5⅜ × 8½. (Available in U.S. only)
21531-8 Pa. $6.95

AMERICAN SAILING SHIPS: Their Plans and History, Charles G. Davis. Photos, construction details of schooners, frigates, clippers, other sailcraft of 18th to early 20th centuries—plus entertaining discourse on design, rigging, nautical lore, much more. 137 black-and-white illustrations. 240pp. 6⅛ × 9¼.
24658-2 Pa. $6.95

ENTERTAINING MATHEMATICAL PUZZLES, Martin Gardner. Selection of author's favorite conundrums involving arithmetic, money, speed, etc., with lively commentary. Complete solutions. 112pp. 5⅜ × 8½. 25211-6 Pa. $2.95

THE WILL TO BELIEVE, HUMAN IMMORTALITY, William James. Two books bound together. Effect of irrational on logical, and arguments for human immortality. 402pp. 5⅜ × 8½. 20291-7 Pa. $7.95

THE HAUNTED MONASTERY and THE CHINESE MAZE MURDERS, Robert Van Gulik. 2 full novels by Van Gulik continue adventures of Judge Dee and his companions. An evil Taoist monastery, seemingly supernatural events; overgrown topiary maze that hides strange crimes. Set in 7th-century China. 27 illustrations. 328pp. 5⅜ × 8½. 23502-5 Pa. $6.95

CELEBRATED CASES OF JUDGE DEE (DEE GOONG AN), translated by Robert Van Gulik. Authentic 18th-century Chinese detective novel; Dee and associates solve three interlocked cases. Led to Van Gulik's own stories with same characters. Extensive introduction. 9 illustrations. 237pp. 5⅜ × 8½.
23337-5 Pa. $4.95

Prices subject to change without notice.

Available at your book dealer or write for free catalog to Dept. GI, Dover Publications, Inc., 31 East 2nd St., Mineola, N.Y. 11501. Dover publishes more than 175 books each year on science, elementary and advanced mathematics, biology, music, art, literary history, social sciences and other areas.